T0321012

NETWORK AND ADAPTIVE SAMPLING

Network
and
Adaptive Sampling

Arijit Chaudhuri
Applied Statistics Unit
Indian Statistical Institute
Kolkata, India

CRC Press
Taylor & Francis Group
Boca Raton   London   New York

CRC Press is an imprint of the
Taylor & Francis Group, an **informa** business

A SCIENCE PUBLISHERS BOOK

CRC Press
Taylor & Francis Group
6000 Broken Sound Parkway NW, Suite 300
Boca Raton, FL 33487-2742

© 2015 by Taylor & Francis Group, LLC
CRC Press is an imprint of Taylor & Francis Group, an Informa business

No claim to original U.S. Government works

Printed on acid-free paper
Version Date: 20140620

International Standard Book Number-13: 978-1-4665-7756-5 (Hardback)

Library of Congress Cataloging-in-Publication Data

Chaudhuri, Arijit, 1940-
    Network and adaptive sampling / Arijit Chaudhuri.
        pages cm
    Summary: "Combining the two statistical techniques of network sampling and adaptive sampling, this book illustrates the advantages of using them in tandem to effectively capture sparsely located elements in unknown pockets. It shows how network sampling is a reliable guide in capturing inaccessible entities through linked auxiliaries. The text also explores how adaptive sampling is strengthened in information content through subsidiary sampling with devices to mitigate unmanageable expanding sample sizes. Empirical data illustrates the applicability of both methods"-- Provided by publisher.
    Includes bibliographical references and index.
    ISBN 978-1-4665-7756-5 (hardback)
    1. Adaptive sampling (Statistics) I. Title.

QA276.6.C4288 2014
519.5'2--dc23   2014018085

Visit the Taylor & Francis Web site at
http://www.taylorandfrancis.com

and the CRC Press Web site at
http://www.crcpress.com

Dedicated

To

Bulu

Preface

Network Sampling was possibly invented by Sirken (1970, 1983) while Chaudhuri & Stenger (2005) briefly narrated his theory; and J.N.K. Rao (1999) further elaborated on this. Chaudhuri's (2000) exposition on it thrives on the foundation laid by Thompson (1990,1992) and Thompson & Seber (1996) on 'Network Sampling', named by the latter two researchers. In this treatise we shall follow this approach.

"Adaptive Sampling" too, from what is understood, originated from the research by Thompson (1990, 1992) extended by Thompson & Seber (1996) and further strengthened by Chaudhuri (2000). Salehi & Seber (2002) and Seber & Salehi (2013) have also contributed immensely to the subject. But the aspects of Network Sampling and Adaptive Sampling which will be discussed in the present volume are confined mainly to the contributions published in the following documents bearing participation by us: Chaudhuri (2000, 2010), Chaudhuri & Saha (2004), Chaudhuri, Bose & Ghosh (2004), Chaudhuri & Stenger (2005), Chaudhuri, Bose & Dihidar (2005) and an exposure by Chaudhuri & Dihidar (2010) plus the current involvement as in Chaudhuri (2011).

First, let us shed some light on the subject. In a standard household survey our intention may be to serviceably estimate the population total or mean of a variable which is an important consideration but is valued zero for many households while it is substantial for many others. However, before conducting the survey one may not have enough data to identify and discriminate these two groups. For example, our interest may be to estimate the mean household expenditure on imparting hospital treatment for an expensive disease to one of the family members. Even if equipped with prior knowledge about the nature of variability of this mean expense territorially in a big area, knocking at probabilistically sampled residential doors may not present a relevant case in point. One is unlikely to be able to gather a list of relevant households with inmates receiving hospital treatment as in-patients facilitating sample selection by sophisticated schemes. An easy way to circumvent this problem is to create a framework for the hospitals/clinics/nursing homes/healthcare centres, choose an appropriate sample therefrom, gather the household addresses of the in-patients who were admitted there during a specific period and contact those households within the geographical area under investigation. Thus two distinct types of units can be identified, i.e., those admitting probability sample selection called Selection Units and the others linked to them as Observation Units with the total number and the initial identifying labels being unavailable. The concern is to estimate the total-mean of the values for the Observation Units from a suitably taken sample of the Selection Units exploiting the "reciprocal link" between the Selection Units and the Observation Units namely, the hospitals (Selection Unit) and households (Observation Unit) providing the in-patients to receive treatment. Between the two "mutually linked" types of units, a "Network" may be established and a sample of Observation Units may be picked up to derive a serviceable estimation procedure. In practice such a "Network Sample" may have a prohibitively large number of Observation Units to be surveyed to ascertain the related values, namely the household expenses on treatment in this particular example. Importantly, it is possible to restrict to a "Constrained Network Sampling", limiting the size to a manageable overall number. The issues can be resolved in what follows.

The theme related to 'Adaptive Sampling' owes its origin to the research made by Thompson (1990, 1992) and Thompson & Seber (1996) and further developed by Chaudhuri (2000). The references cited above concerning 'Network Sampling' except Sirken (1983) and J.N.K. Rao (1999) are also relevant to 'Adaptive Sampling'.

What is the motivating need for this topic? Sometimes our interest focuses on the problem of estimating the total for a variable which takes on positive and substantial values for a moderate number of units forming clusters of such units localized in unknown sites but a zero value on most of the other units in the population. A standard sample design may not capture in such situations enough positive valued units in a sample so drawn to yield a serviceable estimate for the population total. The need then arises for enhancing the information content in the sample which needs to be fruitfully revised in a utilitarian manner. For every unit one may uniquely define a "Neighbourhood" composed of itself plus a few others. One trick may be to leave alone a sampled unit if it bears a zero or a negligible value for the variate of interest; but on the other hand, encountering a positive-valued unit, all its neighbouring units are to be examined for their variate-values, repeating this exercise of finding any positive-valued units in each of the successive neighboring units, stopping the process only when a zero-valued unit is met. All the units thus examined including the starting unit constitute a "Cluster" for the initial unit. The zero-valued units are called the "Edge Units". Omitting the edge units of a cluster of a unit, the leftover set is called a "Network" of the initial unit. Any edge unit or a network composed of a single unit is called a "Singleton Network". Treating them, by courtesy, as genuine "Networks", it follows that the networks are all disjoined and they together coincide with the population of all the units. The union of all the units in the networks of the initial sample of units is called an "Adaptive Sample" as derived from the initial sample. The process resulting in its selection is called "Adaptive Sampling". Obviously it is a device to increase the information content of an initial sample. How serviceable it is in estimation will be discussed in the subsequent sections. It may be mentioned that the total size of an Adaptive Sample may grow exorbitantly higher than the initial sample size. Hence there is a need for a "Constrained Adaptive Sample" equipped with a brake to be applied on an Adaptive sampling process growing beyond control in terms of resources of various kinds. This need is met adequately in this book.

I suppose, the intentions communicated will be easy to comprehend for our readers. Thompson (1990, 1992), Thompson & Seber (1996), Salehi & Seber (2002) and Seber & Salehi (2013) used the phrase "Adaptive Cluster Sampling" with a subtle restriction in connotation compared to that of "Adaptive Sampling" which has also been coined by them. The latter phrase shall be used throughout even though the former one may have been tacitly employed. We beg to be excused if we have inadvertently caused any confusion in the concepts to our readers.

Arijit Chaudhuri
January, 2014   arijitchaudhuri1@rediffmail.com
Kolkata, India   Mobile Number: +919830740074

Acknowledgment

The author gladly expresses his gratitude to the Director of the Indian Statistical Institute for permitting him to continue to work in Indian Statistical Institute after his retirement, as a professor and even after completing 3 years there as a CSIR Emeritus Scientist immediately on superannuation.

He is also obliged to his colleagues in the Applied Statistics Unit of ISI who have maintained a genial atmosphere for creative efforts.

The author
ASU, ISI.

Chapter-wise Summary

CHAPTER 1

Introduction

An important problem in the context of survey sampling arises when many units of a finite population do not bear a characteristic of an investigator's interest while a few with unknown locations do so in plenty. Consequently, a sample at hand no matter howsoever designed, may contain a scanty modicum of relevant units throwing inadequate data to yield accurate estimation of a parameter related to that characteristic.

Two specific techniques are available in the literature to exemplify the information content in such a sample.

CHAPTER 2

Certain Common Methods of Sampling Finite Populations and Estimation Procedures: *Prerequisite for Network and Adaptive Sampling.*

CHAPTER 3

A Plea for Network Sampling with Illustrations Justifying its Use: Theory and Applications.

In order to estimate, for example, the average household expenses for hospital treatment of inmates for specific diseases, a house to house sample survey may not yield enough households of relevance.

A better procedure may be to select a sample of treatment centres, get in touch with the households of the in-patients treated therein a specified period of time and then execute the survey.

A theory of unbiased estimation of a parameter concerned may be easily developed laying down measures of accuracy in estimation.

Actual applications are then illustrated with live data.

CHAPTER 4

Need for Adaptive Sampling: Relevant Theory of Unbiased Estimation

An initially designed sample may not give enough units bearing a characteristic under study. Adaptive sampling is a device to capture more units bearing it retaining unbiasedness in estimation and yielding an enhanced efficiency level.

Theory is developed and application possibilities illustrated.

CHAPTER 5

Adaptive and Network Sampling in Tandem: Illustrating the Advantages

Need for constraining the possibly explosive sample-sizes retaining unbiasedness and maintaining desirable efficiency levels is exposed, development of theories is lucidly presented.

CHAPTER 6

Applications and Case Studies

Maternal mortality, child labour, drug abuses are some of the important problem areas demanding global attention of the social scientists. Some experiences of the author's involvement are narrated illustrating use of emerging statistical methods.

Roughly this publication contains predominantly the author's own contributions with scattered reviews of contemporary works of others.

CHAPTER 7

A Brief Review of Literature

Concepts of Network and Adaptive Sampling by Thompson, Seber and Salehi are first recalled, then the follow-up works by Chaudhuri, Bose, Ghosh, Dihidar, Saha and Pal are briefly recounted.

List of Abbreviations

Actual Coverage Proportion (or Percentage) (ACP)
Applied Statistics Unit (ASU)
Asymptotic Design Consistency (ADC)
Asymptotic Design Unbiasedness (ADU-ness)
Average Coefficient of Variation (ACV)
Average Length (of Confidence interval) (AL)
Central Statistical Organization (now christened "Office"), CSO
Child Labourer (CL)
Coefficients of Variation (CV)
Confidence Coefficient (CC)
Confidence Interval (CI)
Economic Censuses (EC)
Establishments (EST)
Finite Consistency (FC)
First Stage Units (FSU)
Generalized Regression (greg)
Government of India (GOI)
Gross Domestic Products (GDP)
Homogeneous Linear Unbiased Estimators (HLUE)
Horvitz and Thompson's (HT)
Households (HH)
Indian Council of Medical Research (ICMR)

Indian Statistical Institute (ISI)
Injecting Drug Users (IDU)
Last Birthday (LBD)
Maternal Mortality Rate (MMR)
Maximum Likelihood Estimate (MLE)
Mean Square Error (MSE)
Ministry of Social Justice & Empowerment (MSJE),
National Household Survey on Drug Abuses (NHSDA)
National Justice for Social Defence (NJSD)
National Sample Survey Office (NSSO)
National Sample Survey Organizations (now called office NSSO)
Non-Government Organizations (NGOs)
Observation Unit (OU)
Probability Proportional to Size (PPS)
Randomized Response (RR)
Randomized Response Technique (RRT)
Rao-Hartley-Cochran (RHC)
Regional Office for South Asia (ROSA)
Second Stage Units (SSU)
Selection Unit (SU)
Simple Random Sampling (SRS)
Simple Random Sampling With Replacement (SRSWR)
Simple Random Sampling Without Replacement (SRSWOR)
Simple Random Sub-sampling (SRSS)
Simple Random Sub-sampling Without Replacement (WOR)
Statistics and Programme Implementation (MOSPI)
Stone-breaking (SB)
Uni-cluster Designs (UCD)
Uniformly Minimum Variance (UMV)
Union Territories (UT)
United Nations Office on Drug and Crime (UNODC)
West Tripura District (WTD)

Contents

1

Notations and Introduction

1.1 INTRODUCTION

A major problem in survey sampling is that many units of a finite population do not bear characteristics that are being investigated, while a few units bear strong characteristics. Thus, a sample regardless it's design, may contain only a modicum of relevant units resulting in insufficient data to yield accurate estimation of a parameter related to that characteristic.

There are two specific techniques to project the information content in a well planned sample.

One is 'Network Sampling'. We have followed Thompson (1990, 1992) and Thompson & Seber (1996), further extended by Chaudhuri (2000, 2010), in presenting our version of this technique. There are two kinds of units. The first one is 'Selection Units'. These units have a known total number, say M, with identifiable labels so that we may employ a suitable selection scheme or sampling design with a pre-assigned probability to select a sample of Selection Units following a sophisticated procedure.

A second type of unit is called 'Observation Units' conceptually labelled as 1, 2, ..., i, ..., N with N unknown and the Observation Units are not identified and labelled before the start of a survey. Our objective is to estimate the total Y of the values y_i,

$i = 1, \ldots, N$ for a variable of interest defined on the Observation Units respectively, labeled conceptually as $i = 1, \ldots, N$.

The Selection Units and Observation Units in a given context have to be so judiciously defined that one may establish and exploit a link between them so as to be able to choose a sample of Selection Units and through them be able to identify the Observation Units linked to them and moreover gathering further the Selection Units to which the Observation Units thus sampled and identified in addition as 'linked' may enhance the scope of the data content through the 'mutual interlinking' among the Selection Units on the one hand and the Observation Units on the other. Let us elaborate a little more introducing a few notations.

1.2   NOTATIONS AND PRELIMINARIES

Let m_i denote the number of Selection Units linked to the ith Observation Unit $(i = 1, \ldots, N)$. Let $A(j)$ denote the set of Observation Units linked to the jth   Selection Unit $(j = 1, \ldots, M)$. With the objective of unbiasedly estimating the total $Y = \sum_{1}^{N} y_i$, on defining $w_j = \sum_{i \in A(j)} \dfrac{y_i}{m_i}$ and recognizing that

$$W = \sum_{j=1}^{M} w_j \text{ equals } Y$$

one may transfer the problem to one of estimating W. Then choosing a sample s of Selection Units on employing a sampling design p that admits positive values for

$$\pi_j = \sum_{s \ni j} p(s) \quad \text{and} \quad \pi_{jj'} = \sum_{s \ni j, j'} p(s), j, j' \in U = (1, \ldots, j, \ldots, M)$$

one may employ the unbiased estimator

$$e = \sum_{j \in s} \frac{w_j}{\pi_j}$$

for W and hence, for Y too. Then,

$$v = \sum_{j < j' \in s} \sum \left(\frac{\pi_j \pi_{j'} - \pi_{jj'}}{\pi_{jj'}} \right) \left(\frac{w_j}{\pi_j} - \frac{w_{j'}}{\pi_{j'}} \right)^2$$

provides an unbiased estimator for $V(e)$, the variance of e provided $\pi(jj') > 0 \; \forall \; j, j' \in U, (j \neq j')$ every sample contains a common number of distinct units. A serious problem arises if the survey requires too much effort in identifying a very large and relevant set of Observation Units and Selection Units in accomplishing these tasks. Some simple solutions are presented in the later chapters of this book for this problem which is not an insurmountable one.

On the other hand, another kind of situation demands a special type of sampling scheme to be applied. For instance, if a population is composed of a large number of units, many of which are zero-valued in respect of a specific real variable of one's interest while a few of others are positive and even high valued and clustered in unknown locations of concentration. The problem is to suitably estimate the population total of this variable on adopting an appropriate method of sample selection.

In such a situation, "Adaptive Sampling" scheme is often rightly employed where every unit is associated with a well-defined unique "Neighbourhood" composed of itself plus a few more bearing a reciprocal relationship. In case of a situation with a territorial localization, say, for example places with mineral deposits, a plot of land plus four adjacent plots on the right, left, just above and just below of a specifically marked area, each may constitute a neighbourhood. Each plot may contain no mineral deposit or may contain a slight or heavy amounts of such deposits. The total deposit in a province of a country may be needed to be ascertained or estimated. Then one may start with a sample s of units suitably chosen with a selection probability $p(s)$. Anticipating that positive-valued units may be scantily represented in the sample in order to enhance the haul with enough contents, a possibility is to associate with each positive valued unit sampled in all its neighbouring units and repeating this process till positive valued units in the successive neighbourhoods are encountered, stopping only when all the neighbouring units of a unit are zero-valued. The set of units thus reached on starting with an initial unit is called a cluster of the initial unit. Dropping all the zero-valued units in the cluster, one is left with what is called a "Network" of the initial unit. Any zero-valued unit is called an 'Edge Unit'. One may find also networks only of one unit each. Naming these edge units or 'One-unit-clusters' also as Singleton Networks and treating them as genuine Networks it follows that

(i) each network is disjointed, and (ii) all the networks put together exhaust the entire population. Starting with the sample s, say, initially taken if one finishes with observing all the units in the respective networks of all the units sampled, then one reaches a sample called an "Adaptive Sample", say, $A(s)$.

Let $A(i)$ be the network to which a unit i belongs and C_i be its cardinality, i.e. the number of units $A(i)$ contains. Then, it follows that from

$$t_i = 1/C_i \sum_{j \in A(i)} y_j$$

one gets the total $T = \sum_1^N t_i$ and that

$$T = \sum_{i=1}^N 1/C_i \sum_{j \in A(i)} y_j = \sum_{j=1}^N y_j \left[1/C_i \sum_{j \mid A(j) \ni i} 1 \right]$$

$$= \sum_{j=1}^N y_j = Y, \quad \text{since by definition}$$

$$C_i = \sum_{j \mid A(j) \ni i} 1$$

So, to estimate Y it is enough to estimate T.

Hence, $e = \sum_{i \in s} t_i/\pi_i$ is an unbiased estimator of $T = \sum_1^N t_i$ and hence also of Y

and so,

$$V(t) = \sum\sum_{i<j} (\pi_i \pi_j - \pi_{ij}) \left(\frac{t_i}{\pi_i} - \frac{t_j}{\pi_j} \right)^2$$

is the variance of t and an unbiased estimator of this is

$$v(t) = \sum\sum_{i<jts} \left\{ \frac{(\pi_i \pi_j - \pi_{ij})}{\pi_{ij}} \right\} \left(\frac{y_i}{\pi_i} - \frac{y_j}{\pi_j} \right)^2$$

provided every sample contains a common number of district units.

One drawback of Adaptive Sampling is that even if the size of the initial sample s is small or modest, that of the adaptive sample $A(s)$ may turn to be stupendously enormous. So, it is necessary to have an inbuilt device to put a brake in order to have a manageably moderate sample-size. Some devices for this are available in the literature. We shall discuss the procedure in a subsequent chapter.

CHAPTER

2

Sampling and Estimation Methods

Certain Common Methods of Sampling Finite Populations and Estimation Procedures: Pre-Requisite for Network and Adaptive Sampling.

2.1 INTRODUCTORY PRELIMINARIES

We need to refer to a finite survey population $U = (1,\ldots, i, \ldots, N)$ of a known number N of labeled identifiable units, also called "elements" or "individuals". On it are defined real variables y, x, z, w, etc. valued y_i, x_i, z_i, w_i, respectively with totals

$$Y = \sum_i y_i, X = \sum_i x_i, Z = \sum_i z_i, W = \sum_i w_i$$

with respective means

$$\bar{Y} = \frac{Y}{N}, \bar{X} = \frac{X}{N}, \bar{Z} = \frac{Z}{N}, \bar{W} = \frac{W}{N}.$$

The vectors $\underline{Y} = (y_1, \ldots, y_i, \ldots, y_N)$, $\underline{X} = (x_1, \ldots, x_i, \ldots, x_N)$ and $\underline{Z}$, $\underline{W}$ are also similarly defined. The co-ordinates of $\underline{Y}$ are supposed to be totally unknown to begin with, but $\underline{Z}, \underline{W}$ are known vectors while $\underline{X}$ may be fully or partly known, e.g., the total X is often supposed to be known but not the constituent elements. The most frequently addressed problem in survey sampling is to suitably estimate $Y, \overline{Y}$ and $R = \dfrac{Y}{X} = \dfrac{\overline{Y}}{\overline{X}}$ on choosing a sample s of a number of units of U and ascertaining the values of y_i for i in s by dint of a survey. The sample is chosen with a probability $p(s)$ according to a sampling design p to be appropriately employed. We strongly recommend that the reader should consult Chaudhuri (2010).

By $t = t(s, \underline{Y})$ is meant an estimator for Y supposing it to be free of y_j for $j \notin s$. It's performance characteristics are measured by

$$E_p(t) = \sum_s p(s)t(s, \underline{Y}), \text{ it's expectation}$$

$$B_p(t) = E_p(t) - Y, \text{ it's bias in estimating } Y$$

$$M_p(t) = E_p(t - Y)^2, \text{ it's mean square error (MSE) about } Y$$

$$V_p(t) = E_p(t - E_p(t))^2, \text{ it's variance}$$

$$\sigma_p(t) = +\sqrt{V_p(t)}, \text{ it's standard error, and}$$

$$\dfrac{\sigma_p(t)}{t} \times 100 \text{ is the coefficient of variation of } t.$$

It follows that $M_p(t) = V_p(t) + B_p^2(t)$ and t is unbiased for Y if $E_p(t) = Y \ \forall \ \underline{Y}$. It is, vide Chaudhuri (2010), easy to see that

$$\text{Prob}\,[t(s, \underline{Y}) \in (Y - K, Y + K)] \geq 1 - \dfrac{M_p(t)}{K^2}$$

for any positive number K. Consequently, for

$$K = \lambda \sigma_p(t), \lambda > 0, \qquad \text{Prob}\,[t - \lambda\sigma_p(t) \leq Y \leq t + \lambda\sigma_p(t)]$$

$$\geq \left(1 - \dfrac{1}{\lambda^2}\right) - \dfrac{1}{\lambda^2}\left(\dfrac{|B_p(t)|}{\sigma_p(t)}\right)^2$$

So, for an unbiased estimator t for Y, one gets a Confidence Interval for Y as $(t - \lambda\sigma_p(t), t + \lambda\sigma_p(t))$ with a Confidence Coefficient at least as high as $\left(1 - \dfrac{1}{\lambda^2}\right)$ which is $\dfrac{8}{9}$ if one takes $\lambda = 3$. So, it is desirable to employ for Y an unbiased estimator with as small a standard error as possible. It is important to observe however that

1. For the existence of an unbiased estimator for Y, a 'necessary and sufficient condition' is that the design p must have a 'uniformly positive inclusion probability' π_i for every i in U, writing

$$\pi_i = \sum_{s \ni i} p(s) \quad \text{[vide Chaudhuri (2010)]}$$

2. No design p, unless it is a Census Design, admits an unbiased estimator with its variance uniformly the least for every $\underline{Y}$ in, $\Omega = \{\underline{Y}| - \alpha < a_i \le Y_i \le b_i < + \alpha, i = 1,\dots,N\}$ as has been shown by Basu (1971).

    A census design p is one with $p(s) = 0$, unless s contains all the units of U. Godambe (1955) introduced the class of Homogeneous Linear Unbiased Estimators for Y of the form

$$t_b = \sum_{i \in s} y_i b_{si}$$

with b_{si}'s as quantities free of $\underline{Y}$ subject to

$$\sum_{s \ni i} p(s) b_{si} = 1 \forall i \in U$$

but he proved the "Non-existence theorem":

3. For a general design p, among Homogeneous Linear Unbiased Estimators for Y there does not exist one with the Uniformly Minimum Variance property.

For an exceptional class of Uni-cluster Designs for which given any two samples s_1 and s_2 with $p(s_1) > 0$, $p(s_2) > 0$, either (i) $s_1 \cap s_2 = \phi$, the empty set or (ii) $s_1 \sim s_2$, i.e., any unit in s_1 is in s_2 and vice versa, however, Godambe's negative result (3) does not hold but the Horvitz & Thompson's (1952) estimator

$$t_{HT} = \sum_{i \in s} \frac{y_i}{\pi_i}$$

is of course an Homogeneous Linear Unbiased Estimators for Y and has the uniformly smallest variance if the design p is a Uni-cluster Design.

This result given by Hege (1965), Hanurav (1966) and Lanke (1975) is a consequence of the findings by Basu (1958) and by Basu and Ghosh (1967) as explained below in brief. We shall throughout restrict the study only to non-informative designs for which $p(\sigma)$ is free of elements of $\underline{Y}$.

For a given sample $\sigma = (i_1,\ldots, i_n)$ of n units of U arranged in a successive order, it does not matter if any one is repeated with several multiplicities, by $d = (\sigma, y_i \mid i \in \sigma)$, we mean the survey data.

Corresponding to the ordered sample s of units permitted to have varying multiplicities, let $\sigma^* = \{j_1,\ldots,j_k\}$, $1 \le k \le n$, denote the unordered sample of k distinct units each of which is one of the elements in σ. Also, let $d^* = \{\sigma^*, y_j \mid j \in \sigma^*\}$. It then follows that given the data point d, the data point d^* is a (1) 'Sufficient' statistic and moreover, it is (2) the 'Minimal Sufficient' statistic. Thus, (1)' Prob $(d|d^*)$ is free of $\underline{Y}$ and (2)', given any other sufficient statistic $t(d)$, given d, $t(d)$ is a function of d^* i.e., d^* is (2), the minimal sufficient statistic.

A crucial consequence of this is:

Given a statistic $t = t(\sigma, \underline{Y})$, let a derived statistic be

$$t^*(\sigma^*) = \frac{\sum_{\sigma \to \sigma^*} p(\sigma) t(\sigma, \underline{Y})}{\sum_{\sigma \to \sigma^*} p(\sigma)},$$

the symbol $\sum_{\sigma \to \sigma^*}$ denoting the sum over the samples to each of which corresponds the same σ^* as the set of units with ordering ignored and the multiplicity of each unit in σ^* being unity. Then

$$E_p(t) = E_p(t^*)$$

and

$$V_p(t^*) = V_p(t) - E_p(t - t*)^2.$$

Thus

$$V_p(t^*) \le V_p(t)$$

and

$$V_p(t^*) < V_p(t) \quad \text{if} \quad \text{Prob}(t \ne t^*) = 1.$$

This gives us a 'Complete Class Theorem in Survey Sampling'.

2.2 SAMPLING AND ESTIMATION METHODS

Let us consider a few schemes of sample selection and methods of estimating $Y, \bar{Y}, R$ based on samples accordingly drawn.

1. Simple Random Sampling (SRS) With Replacement (SRSWR).

Here one fixes a number of draws, say n, and on each draw using a table of random numbers selects 1 unit of U with a probability $1/N$ and repeats this independently n times.

Writing Y_r as the value of y for the unit chosen on the rth draw $(r = 1,\ldots, n)$,

$$\bar{y} = \frac{1}{n}\sum_{r=1}^{n} Y_r,$$

the sample mean based on all the n draws, has $E_p(\bar{y}) = \bar{Y}$, with a variance

$$V_p(\bar{y}) = \frac{\sigma^2}{n},$$

writing

$$\sigma^2 = \frac{1}{N}\sum_{1}^{N}(y_i - \bar{Y})^2,$$

the variance of all the values $y_1, \ldots, y_N$.
Writing

$$s^2 = \frac{1}{n-1}\sum_{r=1}^{n}(Y_r - \bar{y})^2,$$

an unbiased estimator for $V_p(\bar{y})$ is $\dfrac{s^2}{n}$.

By $CV = 100\dfrac{\sigma}{\bar{Y}}$ we shall mean the Coefficient of Variation of the population values $y_1, \ldots, y_N$.

By $CV(t) = 100\dfrac{\hat{\sigma}_p(t)}{t}$, we shall mean the estimated Coefficient of Variation of an estimator t, writing $\hat{\sigma}_p(t)$ for an estimator of its standard error.

Based on experience an expert thumb rule says that if

1. $CV(t) \leq 10\%$, t is an excellent estimator for Y
2. $10\% < CV(t) \leq 20\%$, t is a good estimator

   3. $20\% < CV(t) \leq 30\%$, t is acceptable

   4. $CV(t) > 30\%$, t need not be used.

    Basu (1958), invoking the principle of sufficiency, showed that based on Simple Random Sampling With Replacement, the sample mean $\bar{y}_d$ based on the d distinct units out of a total of n units drawn is also unbiased for $\bar{Y}$ and has a variance smaller than $V_p(\bar{y})$.

    For Simple Random Sampling With Replacement in n draws, the inclusion-probability

$$\pi_i = 1 - \left(\frac{N-1}{N}\right)^n \quad \text{for every } i = 1, \ldots, N$$

and the inclusion-probability

$$\pi_{ij} = \sum_{s \ni i,j} p(s) \quad \text{for } i, j = 1, \ldots, N \ (i \neq j)$$

for Simple Random Sampling With Replacement works out as

$$\pi_{ij} = 1 - 2\left(\frac{N-1}{N}\right)^n + \left(\frac{N-2}{N}\right)^n$$

which has been derived utilizing De Morgan's law because writing A_i = the event that $s \ni i$, A_j the event that

$$s \ni j, \text{ Prob } [A_i \cap A_j] = \text{Prob}[(A_i^c \cup A_j^c)^c].$$

Writing
$$I_{si} = 1 \text{ if } i \in s$$
$$= 0 \text{ if } i \notin s,$$

and
$$I_{sij} = I_{si} I_{sj},$$

it follows that based on a design p admitting $\pi_i > 0 \ \forall \ i$, variance of the Horvitz & Thompson's estimator $t_{HT} = \sum_{i \in s} \dfrac{y_i}{\pi_i}$ is

$$V(t_{HT}) = \sum y_i^2 \frac{1 - \pi_i}{\pi_i} + \sum\sum_{i \neq j} y_i y_j \frac{\pi_{ij} - \pi_i \pi_j}{\pi_i \pi_j}$$

as given by Horvitz & Thompson (1952) and an unbiased estimator for it is

$$v_{HT}(t_{HT}) = \sum y_i^2 \frac{1-\pi_i}{\pi_i} \frac{I_{si}}{\pi_i} + \sum\sum_{i\neq j} y_i y_j \frac{\pi_{ij}-\pi_i\pi_j}{\pi_i\pi_j} \frac{I_{sij}}{\pi_{ij}}$$

provided $\pi_{ij} > 0 \;\forall\; i \neq j \in U$.

Chaudhuri & Pal (2002), on modifying the works of Hajek (1958) and J.N.K. Rao (1979) have derived the following alternative to $v_{HT}(t_{HT})$:

$$v_{CP}(t_{HT}) = \sum\sum_{i<j} \left(\frac{y_i}{\pi_i}-\frac{y_j}{\pi_j}\right)^2 (\pi_i\pi_j-\pi_{ij})\frac{I_{sij}}{\pi_{ij}} + \sum \frac{y_i^2}{\pi_i}\beta_i\frac{I_{si}}{\pi_i}$$

writing

$$\beta_i = 1 + \frac{1}{\pi_i}\sum_{j\neq i}^N \pi_{ij} - \sum_1^N \pi_i$$

In case every s with $p(s) > 0$ has a common number of distinct units, β_i equals zero and in that case

$$v_{YG}(t_{HT}) = \sum\sum_{i<j} \left(\frac{y_i}{\pi_i}-\frac{y_j}{\pi_j}\right)^2 \left(\frac{\pi_i\pi_j-\pi_{ij}}{\pi_{ij}}\right)\frac{I_{sij}}{\pi_{ij}}$$

is an alternative unbiased variance estimator given much earlier by Yates & Grundy (1953).

For Simple Random Sampling with replacement obviously $\beta_i \neq 0$ and both $v_{HT}(t_{HT})$ and $v_{CP}(t_{HT})$ may be employed in deriving $CV(t_{HT})$.

2. Simple Random Sampling Without Replacement (SRSWOR)

It leads to an unordered sample of distinct units, say, n ($2 \leq n < N$) in number.

For this

$$\pi_i = \frac{n}{N}\forall i \in U$$

and

$$\pi_{ij} = \frac{n(n-1)}{N(N-1)}\forall i \neq j \text{ in } U.$$

For $\overline{Y}$ the sample mean $\overline{y} = \frac{1}{n}\sum_{i\in s} y_i$ is an unbiased estimator with a variance

$$V_p(\bar{y}) = \frac{N-n}{N}\frac{S^2}{n},$$

writing

$$S^2 = \frac{1}{(N-1)}\sum_{i=1}^{N}(y_i - \bar{Y})^2.$$

Writing

$$s^2 = \frac{1}{(n-1)}\sum_{i\in s}(y_i - \bar{y})^2 \text{ one has}$$

$$v_{\text{SRSWOR}}(\bar{y}) = \frac{N-n}{Nn}s^2 \text{ as its unbiased estimator for}$$

$$V_p(\bar{y}) = V_{(\text{SRSWOR})}(\bar{y}) = \frac{N-n}{Nn}s^2.$$

An estimated Coefficient of Variation of $\bar{y}$ is then

$$(\bar{y}) = 100\frac{s}{\bar{y}}\sqrt{\frac{N-n}{Nn}}.$$

An interesting problem in the context of simple random sample without replacement is determining the appropriate sample-size n, given the population size N. In order to settle this issue one needs to specify (i) how much error $|\bar{y} - \bar{Y}|$ in estimating $\bar{Y}$ may be tolerated with how much (ii) Probability and (iii) how much heterogeneity level is anticipated inherently in the unknown values in the population $(U, \underline{Y}) = ((1, y_1), \ldots, (i, y_i), \ldots, (N, y_N)$.

Thus, for example, we may analytically formulate

$$\text{Prob}\,[|\,\bar{y} - \bar{Y}\,| < \theta\bar{Y}] \geq 1 - \frac{V(\bar{y})}{(\theta\bar{Y})^2},$$

with θ as a small positive proper fraction as, say, 1/10 .

Then

$$1 - \frac{V(\bar{Y})}{(\theta\bar{Y})^2} = 1 - 100\left[\frac{CV(\bar{y})}{100}\right]^2 = \text{confidence coefficient} = 1 - \alpha.$$

Setting α at 0.05 and $CV(\bar{y})$ at various levels we may tabulate n against $(N,\alpha,CV(\bar{y}))$.

Another intriguing problem dealt with by Freund (1994) gives an unbiased estimator as follows for N, somehow supposed to be forgotten, utilizing the realized labels in a Simple Random Sample Without Replacement of size n drawn from a survey population of size N.

Supposing that X is the largest of the realized labels, one may check that

$$\text{Prob}(X = x) = \frac{\binom{x-1}{n-1}}{\binom{N}{n}}, x = n, n+1, \ldots, N$$

giving

$$1 = \frac{1}{\binom{N}{n}} \sum_{x=n}^{N} \binom{x-1}{n-1} \Rightarrow \sum_{x=n}^{N} \binom{x-1}{n-1} = \binom{N}{n}$$

and

$$E(X) = \frac{1}{\binom{N}{n}} \sum_{x=n}^{N} x \binom{x-1}{n-1} = n\frac{\binom{N+1}{n+1}}{\binom{N}{n}} = \frac{n(N+1)}{(n+1)}$$

Hence, $\hat{N} = \dfrac{(n+1)x}{n} - 1$ is an unbiased estimator for N, writing x for the largest realized label. It may further be checked that

$$V_{\text{SRSWOR}}(X) = E\,X(X+1) - E(X) - [E(X)]^2$$

$$= \frac{n(n+1)\binom{N+2}{n+2}}{\binom{N}{n}} - \frac{n}{n+1}(N+1) - \left[\frac{n(N+1)}{n+1}\right]^2$$

$$= \frac{(N+1)}{(n+1)^2}\left[\frac{(N+2)n(n+1)^2}{(n+2)} - n(n+1) - n^2(N+1)\right]$$

and $CV_{\text{SRWSOR}}(X)$ follows at once.

3. Stratified sampling

Population U is split up into a number of mutually exclusive components called *strata* of respective sizes $N_1, \ldots, N_h, \ldots, N_H$, so that

$$\sum_{h=1}^{H} N_h = N.$$

From each respective stratum then a sample is taken independently to yield a 'stratified sample'. If only Simple Random Sample Without Replacements of respective sizes $n_h \left(h = 1, .., H, \sum_h n_h = n \right)$ are taken from the respective strata, then for $\bar{Y} = \sum W_h \bar{Y}_h \equiv$ the population mean, writing $W_h = \dfrac{N_n}{N}$ and $\bar{Y}_h = \dfrac{1}{N_h} \sum_{i=1}^{N_h} y_{hi}$, the strata-means and $y_{hi}(i = 1, \ldots, N_n; h = 1, \ldots, H)$ the values of y for the ith unit in the hth stratum an unbiased estimator is taken as

$$\bar{y}_{st} = \sum_h W_h \bar{y}_h,$$

$$\bar{y}_h = \frac{1}{n_h} \sum_{i \in s'_h} y_{hi},$$

the sample-mean of the hth stratum, of which s'_h is a sample.

For such a 'stratified Simple Random Sampling Without Replacement'

$$V(\bar{y}_{st}) = \frac{1}{N^2} \sum_{h=1}^{H} N_h (N_h - n_h) \frac{S_h^2}{n_h}$$

writing

$$S_h^2 = \frac{1}{N_h - 1} \sum_{1}^{N_h} (y_{hi} - \bar{Y}_h)^2 .$$

Since

$$\left(\sum_{h=1}^{H} n_h \right) \left(\sum \frac{N_h^2 S_h^2}{n_h} \right) \geq \left(\Sigma N_h S_h \right)^2,$$

the optimal choice of n_h, given n is

$$n_h(opt) = \frac{nN_hS_h}{\Sigma N_hS_h}$$

Using this allocation the optimal variance is

$$V_{opt} = \frac{(\Sigma N_hS_h)^2}{Nn} - \frac{\Sigma N_hS_h^2}{N^2}.$$

Letting $s_h^2 = \frac{1}{(n_h - 1)}\sum_{i \in s_h}(y_{hi} - \bar{y}_h)^2$, an unbiased estimator for $V(\bar{y}_{st})$ is

$$v_{st}(\bar{y}_{st}) = \frac{1}{N^2}\sum N_h(N_h - n_h)\frac{s_h^2}{n_h}.$$

4. Probability proportional to size (PPS) sampling

On $U = (1, \ldots i, \ldots, N)$, suppose $\underline{Y}$ and $\underline{X}$ are defined with $x_i > 0 \; \forall \; i \in U$.

Let $p_i = \frac{x_i}{X}, i \in U$, called the 'Normed size – measures', be known and x_i's be the integers. If not, let them all be made so multiplying each by 10^K, with K suitably chosen.

Lahiri (1951) gave us the Probability Proportional to Size method to choose a unit from U in the following manner:

Take a random number i between 1 and N and independently another random number R between 1 and M on choosing the number $M \geq \frac{\max\limits_{1 \leq i \leq N} (x_i)}{}$.

Then, if $R \leq x_i$, take i into the sample and if $> x_i$, then drop (i, R) and repeat the process till the unit i is eventually taken into the sample. Noting that

$$Q = 1 - \frac{1}{N}\frac{1}{M}\sum_{i=1}^{N} x_i$$

is the probability that according to this procedure no unit is selected into the sample, the probability of choosing i is

$$\text{Prob}(i) = \frac{1}{N}\frac{x_i}{M} + Q\left(\frac{1}{N}\frac{x_i}{M}\right) + Q^2\left(\frac{1}{N}\frac{x_i}{M}\right) + \cdots$$

$$= \frac{1}{N}\frac{x_i}{M}\left(1 + Q + Q^2 + \cdots\right) = \frac{1}{N}\frac{x_i}{M}\left(1 - Q\right)^{-1}$$

$$= \frac{1}{N}\frac{x_i}{M}\left(\frac{M}{\overline{X}}\right) = \frac{x_i}{N\overline{X}} = p_i.$$

In order to estimate Y, let n draws be made with replacement on each draw choosing an individual into the sample by the above scheme of Lahiri. Then, writing P_r as the value of p_i for the unit selected on the rth draw and Y_r as the value of y chosen on the rth, draw an unbiased estimator for Y as

$$t_{HH} = \frac{1}{n}\sum_{r=1}^{n}\frac{Y_r}{P_r} \text{ because } E\left(\frac{Y_r}{P_r}\right) = \sum_{1}^{N}\frac{y_i}{p_i}p_i = Y \text{ for every } r = 1, ..., n.$$

Also,

$$V\left(\frac{Y_r}{P_r}\right) = \sum \frac{y_i^2}{p_i} - Y^2 = \sum_{1}^{N}p_i\left(\frac{y_i}{p_i} - Y\right)^2$$

$$= \sum\sum_{i<j}p_i p_j\left(\frac{y_i}{p_i} - \frac{y_j}{p_j}\right)^2 = V, \text{ say,}$$

so that

$$V(t_{HH}) = \frac{V}{n}.$$

Since for

$$r \neq r'(=1,...,n), \ E\left(\frac{Y_r}{P_r} - \frac{Y_{r'}}{P_{r'}}\right)^2 = E\left[\left(\frac{Y_r}{P_r} - Y\right) - \left(\frac{Y_{r'}}{P_{r'}} - Y\right)\right]^2 = 2V,$$

it follows that

$$v = \frac{1}{2n^2(n-1)}\sum_{r\neq r'}^{n}\sum_{}^{n}\left(\frac{Y_r}{P_r} - \frac{Y_{r'}}{P_{r'}}\right)^2$$

is an unbiased estimator for the variance of t_{HH}, the estimator for Y due to Hansen and Hurwitz (1943).

5. Rao, Hartley and Cochran's (1962) method of sample selection and estimation of Y

To choose a sample of n distinct units from U, first n random groups are formed. To do so, positive integers N_1, N_2, ..., N_n are chosen with their sum $\sum_n N_i$ equal to N. Then, first a Simple Random Sample Without Replacement of N_1 units is chosen from N units of U, then from the $N - N_1$ units a Simple Random Sample Without Replacement of N_2 units is drawn and so on and finally a Simple Random Sample Without Replacement of N_n units is taken from the remaining $N - N_1 - N_2, \cdots - N_{n-1}$ units of U.

The normed positive numbers

$$p_1,\ldots,p_N \left(0 < p_i < 1, \sum_1^N p_i = 1 \right)$$

are known to be respectively associated with the units $i(= 1, \ldots, N)$ of the population $U = (1, \ldots, N)$.

Let $i_1,\ldots, i_j, \ldots, i_{N_i}$ be the N_i units in the ith group thus formed bearing $p_{i1},\ldots,p_{ij},\ldots,p_{iN_i}$ as their normed size-measures. Let $Q_i = p_{i1} + \cdots + p_{iN_i}$. Now from each of these groups just one is selected independently of each other with the respective probabilities $p_{ij}/Q_i, j = 1, \ldots, N_i, i = 1, \ldots, n$.

Then, Rao, Hartley and Cochran's (1962) unbiased estimator for Y is

$$t_{RHC} = \sum_n y_i \frac{Q_i}{p_i},$$

writing (y_i, p_i) for the (y, p)-value of the unit chosen from the ith group.

Writing

$$A = \frac{\sum_n N_i^2 - N}{N(N - 1)}$$

and

$$B = \frac{\sum_n N_i^2 - N}{N^2 - \sum_i N_i^2},$$

their formulae are

$$V(t_{RHC}) = A \sum \sum_{i<j} p_i p_j \left(\frac{y_i}{p_i} - \frac{y_j}{p_j} \right)^2$$

and

$$v(t_{RHC}) = B \sum_n \sum_n Q_i Q_j \left(\frac{y_i}{p_i} - \frac{y_j}{p_j} \right)^2$$

writing $\sum_n \sum_n$ as the sum over the distinct pairs of the groups with no repetition, for the variance of t_{RHC} and of an unbiased estimator thereof.

Rao, Hartley and Cochran (1962) have given the optimal formulae yielding the minimal value of $V(t_{RHC})$ as $N_i = \left[\frac{N}{n} \right]$ for $i = 1, \ldots, K$ and $= \left[\frac{N}{n} \right] + 1$ for $i = K + 1, \ldots, n$ with K uniquely determined giving $\sum_n N_i = N$.

6. M.N. Murthy's (1957) sampling strategy

Let

$$\underline{p} = (p_1, \ldots, p_i, \ldots, p_N), \left(0 < p_i < 1 \forall i, \sum_1^N p_i = 1 \right)$$

be the known positive normed size-measures, associated with the units of $U = (1, \ldots, i, \ldots, N)$. Taking any sampling design p involving the entries in $\underline{p}$ with $p(\delta)$ as the selection probability of a sample δ of n units, each distinct, and $p(\delta|i)$ as the conditional selection-probability of δ, of which i is the unit chosen on the 1st draw with probability p_i, Murthy's (1957) unbiased estimator for Y is

$$t_M = \frac{1}{p(\delta)} \sum_{i \in s} y_i p(\delta | i)$$

Clearly,

$$p(\delta) = \sum_{i \in s} p_i p(\delta | i) .$$

A sampling method plus an estimation procedure is called a "Sampling Strategy".

A formula for $V(t_M)$ follows on noting the results of Ha'jek (1958) and J.N.K. Rao (1979), as elaborated by Chaudhuri (2010).

Let $t_b = \sum_{i=1}^{N} y_i b_{si} I_{si}$ with b_{si}'s as constants free of $\underline{Y} = (y_1, ..., y_N)$.

Then the Mean Square Error of t_b about Y is

$$M = E_p(t_b - Y)^2 = \sum_i \sum_j y_i y_j d_{ij}$$

writing

$$d_{ij} = E_p(b_{si} I_{si} - 1)(b_{sj} I_{sj} - 1).$$

Let

$$w_i \neq 0 \,\forall\, i \in U, z_i = \frac{y_i}{w_i}, \, i \in U.$$

Then

$$M = \Sigma\Sigma z_i z_j w_i w_j d_{ij}.$$

Let

$$\alpha_i = \sum_{j=1}^{N} d_{ij} w_j.$$

Ha'jek (1958) and J.N.K. Rao (1979) observed that if a Non-Negative Definite Quadratic Form (NNDQF) $Q = \sum_{i=1}^{T} \sum_{j=1}^{T} a_{ij} x_i x_j$ has

$\sum_{i=1}^{T} \sum_{j=1}^{T} a_{ij} = 0$, then, it follows that $\sum_{j=1}^{T} a_{ij} = 0 \,\forall\, i = 1, .., T$. Utilizing

these Chaudhuri & Pal (2002) showed the following:

$$M = -\sum_{i<j} \sum w_i w_j \left(\frac{y_i}{w_i} - \frac{y_j}{w_j}\right)^2 d_{ij} + \sum \frac{y_i^2}{w_i} \alpha_i$$

In case $w_i = c \,\forall\, i$ as it holds in many specific forms of t_b including t_M, then, Ha'jek (1958) and J.N.K. Rao (1979) noted that

$$M = -\sum_{i<j} \sum w_i w_j \left(\frac{y_i}{w_i} - \frac{y_j}{w_j}\right)^2 d_{ij},$$

as also noted by Chaudhuri (2010).

From this it follows that

$$V(t_M) = \sum\sum_{i<j} a_{ij} \left[1 - \sum_{s \ni ij} \frac{p(s\,|\,i)p(s\,|\,j)}{p(s)} \right]$$

on writing

$$a_{ij} = p_i p_j \left(\frac{y_i}{p_i} - \frac{y_j}{p_j} \right)^2.$$

Further it is easy to check that

$$\hat{M} = \sum\sum_{i<j} a_{ij} \frac{I_{sij}}{p^2(s)} [p(s\,|\,i,j)p(s) - p(s\,|\,i)p(s\,|\,j)]$$

is an unbiased estimator of $V(t_M)$.

7. Lahiri (1951), Midzuno (1952), Sen's (1953) scheme of sampling and unbiased ratio estimation

On the first draw from U, a unit is drawn with the probability p_i for i in U and this is followed by choosing a Simple Random Sample Without Replacement of $(n - 1)$ units from the remaining $(N - 1)$ units of the population U. Subsequently, the selection-probability of the sample s of these n units drawn is

$$p(s) = \sum_{i \in s} \left[p_i \frac{1}{\binom{N-1}{n-1}} \right] = \frac{\sum\limits_{i \in s} p_i}{\binom{N-1}{n-1}} = \frac{\sum\limits_{i \in s} x_i}{X \binom{N-1}{n-1}}$$

taking $\quad p_i = \dfrac{x_i}{X}, x_i > 0, \ X = \sum_1^N x_i.$

The ratio estimator

$$t_R = X \frac{\sum\limits_{i \in s} y_i}{\sum\limits_{i \in s} x_i} = X \frac{\bar{y}}{\bar{x}}$$

is unbiased for Y because

$$E_p(t_R) = \sum_s \left[\left(X \frac{\sum\limits_{i\in s} y_i}{\sum\limits_{i\in s} x_i} \right) \left(\frac{\sum\limits_{i\in s} x_i}{X\binom{N-1}{n-1}} \right) \right] = \frac{1}{\binom{N-1}{n-1}} \left(\sum_{i=1}^N y_i \right) \left(\sum_{s\ni i} 1 \right) = Y$$

For this scheme of sample selection,

$$p(s \mid i) = \frac{1}{\binom{N-1}{n-1}}$$

and

$$p(s \mid i, j) = \frac{1}{\binom{N-2}{n-2}}, \; i, j \in U.$$

So the exact variance of the ratio estimator $t_R = X\dfrac{\bar{y}}{\bar{x}}$ is

$$V(t_R) = \sum_{i<j}\sum x_i x_j \left(\frac{y_i}{x_i} - \frac{y_j}{x_j} \right)^2 \left[1 - \frac{1}{\binom{N-1}{n-1}} X \sum_{s\ni i,j} \left(\sum_{i\in s} x_i \right) \right].$$

An exactly unbiased estimator of this $V(t_R)$ using the results in the previous section is

$$v(t_R) = \sum_{i<j}\sum p_i p_j \left(\frac{y_i}{p_i} - \frac{y_j}{p_j} \right)^2 \left[1 - \frac{1}{\binom{N-1}{n-1}} \sum_{s\ni i,j} \left(\sum_{i\in s} p_i \right) \right].$$

For this Lahiri, Midzuno, Sen (LMS) scheme,

$$\pi_i = p_i + (1 - p_i)\frac{n-1}{N-1}$$

and

$$\pi_{ij} = p_i \left(\frac{n-1}{N-1} \right) + p_j \left(\frac{n-1}{N-1} \right) + (1 - p_i - p_j)\frac{(n-1)(n-2)}{(N-1)(N-2)}.$$

Let t be an unbiased estimator for Y. Then, $V_p(t) = E_p(t^2) - Y^2$. This immediately gives

$$v_p(t) = t^2 - \left[\sum y_i^2 \frac{I_{si}}{\pi_i} + \sum\sum_{i \neq j} y_i y_j \frac{I_{sij}}{\pi_{ij}} \right]$$

as an unbiased estimator for $V_p(t)$. For the ratio estimator t_R based on the Lahiri, Midzuno and Sen's scheme of sampling, an alternative unbiased estimator for $V_p(t_R)$ immediately follows.

8. Hartley–Ross ratio-type unbiased estimator based on a Simple Random Sample Without Replacement of size n.

Since covariance between $r_i = \dfrac{y_i}{x_i}$ and x_i is

$$C = \frac{1}{N} \sum_1^N \left(r_i - \frac{1}{N} \sum_1^N r_i \right)(x_i - \bar{X}) = \bar{Y} - \bar{X}\bar{R},$$

writing

$$\bar{R} = \frac{1}{N} \sum_1^N r_i = \frac{1}{N} \sum_1^N \frac{y_i}{x_i},$$

then,

$$\bar{r} = \frac{1}{n} \sum_{i \in s} r_i$$

is an unbiased estimator for $\bar{R}$. Assuming as usual that x_i, $i \in U$ are all known and observing that

$$\hat{C} = \left(\frac{N-1}{N} \right) \frac{1}{n-1} \sum_{i \in s}(r_i - \bar{r})(x_i - \bar{x}) = \frac{(N-1)n}{N(n-1)}(\bar{y} - \bar{r}\,\bar{x})$$

unbiasedly estimates C, an unbiased estimator of Y is

$$\hat{Y}_{HR} = X\bar{r} + \frac{(N-1)n}{(n-1)}(\bar{y} - \bar{r}\,\bar{x}).$$

An obvious unbiased estimator for $V_{\mathrm{SRSWOR}}(\hat{Y}_{HR})$ is

$$v_1\left(\hat{Y}_{HR}\right) = (\hat{Y}_{HR})^2 - \left[\frac{N}{n}\sum_{i\in s}y_i^2 + \frac{N(N-1)}{n(n-1)}\sum_{i\neq j\in s}\sum y_i y_j\right]$$

and another one is

$$v_2\left(\hat{Y}_{HR}\right) = (\hat{Y}_{HR})^2 - (2N-1)\frac{N}{n}\sum y_i^2 + \frac{N(N-1)}{n(n-1)}\sum_{i\neq j}\sum (y_i - y_j)^2$$

because

$$Y^2 = \sum_1^N y_i^2 + \sum_{i\neq j}\sum y_i y_j = (2N-1)\sum_1^N y_i^2 - \sum_{i\neq j}\sum (y_i - y_j)^2.$$

3

Plea for Network Sampling

A plea for Network Sampling with Illustrations Justifying Its Use: Theory and Applications.

3.1 Introduction

In order to estimate, for example, the average household expenses for hospital treatment of inmates for specific illnesses a house-to-house sample survey may not fetch enough Households of relevance, i.e. those with members receiving such treatments. A better procedure may be to select a sample of treatment centres, get in touch with the Households of the in-patients treated there in a specified period of time and then execute the survey.

A theory of unbiased estimation of a parameter concerned may then be easily developed laying down measures of accuracy in estimation.

Actual applications are then illustrated with live data of our own involvements.

3.2   NETWORK SAMPLING AND CONSTRAINED NETWORK SAMPLING PROCEDURES IN ONE OR MORE STAGES

3.2.1   With the Background thus Far Spelled Out It Seems Proper We may Straightaway Develop the Theory as Follows in Brief for Estimating the Survey Population Total of a Variable of Interest

Let y be a real-valued variable of interest taking values y_i on our "Observational Unit" conceptually labeled i over a finite number N of unidentified units $i = 1, \ldots, N$. The number N is unknown and we suppose that we cannot start with a list of these units. But these Observational Units are supposed to be "linked" conceptually with an identifiable set of a known number M of "Selection Units" in a well-defined manner. Our object is to unbiasedly estimate

$$Y = \sum_1^N y_i$$ and get a proper measure of accuracy in estimation.

Let m_i be the number of Selection Units to which the ith Observational Unit is linked and $A(j)$ be the set of Observational Units linked to the jth Selection Unit ($j = 1, \ldots, M$; $i = 1, \ldots, N$). Our intention is to suitably estimate the total

$$Y = \sum_{i=1}^N y_i$$ on surveying the Observational Unit's linked

to a suitably chosen sample s of Selection Units with a probability $p(s)$, according to an appropriately hit upon sampling design p. On contacting the selected sample s of Selection Units, the Observational Units linked to them are contacted and the other listed Selection Units to which the respective Observational Units are identified to be linked are also identified provided the Selection Units are in the sample s. Thus the 'links' existing among the Observational Units with the sampled Selection Units on the one hand and those among the sampled Selection Units with as many Observational Units reciprocally linked are exploited to the extent possible. The complete set of the Observational Units thus encountered constitutes a "Network Sample".

Theorem 1

$$w_j = \sum_{i \in A(j)} \frac{y_i}{m_i} \text{ yields } W = \sum_{j=1}^{M} w_j \text{ which equals } Y = \sum_{i=1}^{N} y_i.$$

Proof:

$$W = \sum_{j=1}^{M} w_j = \sum_{j=1}^{M} \left(\sum_{i \in A(j)} \frac{y_i}{m_i} \right)$$

$$= \sum_{i=1}^{N} y_i \left(\frac{1}{m_i} \sum_{j|A(j) \ni i} 1 \right)$$

$$= \sum_{1}^{N} y_i = Y$$

because $\displaystyle\sum_{j|A(j) \ni i} 1$ equals m_i by definition.

Theorem 2

$t_b = \displaystyle\sum_{j \in s} w_j b_{sj}$ with b_{sj} for $j \in s$ for every s with $p(s) > 0$ as a constant free of $\underline{W} = (w_1, \ldots, w_j, \ldots, w_M)$ such that $\displaystyle\sum_{s \ni j} p(s) b_{sj} = 1$ for every $j = 1, \ldots, M$ satisfies $E_p(t_b) = W = Y$, i.e. t_b is an unbiased estimator for Y.

Proof: Easy and hence omitted.

Letting

$$d_{jj'} = E_p(b_{sj} I_{sj} - 1)(b_{sj'} I_{sj'} - 1),$$

$$I_{sj} = 1 | 0 \text{ as } s \ni j | s \not\ni j$$

and

$$I_{sjj'} = 1 | 0 \text{ as } s \ni (j, j') | s \not\ni (j, j'), \Rightarrow I_{sjj'} = I_s I_{sj'}.$$

$$V_p(t_b) = \sum_{j=1}^{M} \sum_{j'=1}^{M} w_j w_{j'} d_{jj'}.$$

Let there exist constants $a_j \neq 0$, $j = 1, \ldots, M$.

Then, let $\alpha_j = \displaystyle\sum_{j'=1}^{M} d_{jj'} a_j$. Alternatively,

$$V_p(t_b) = -\sum_{j=1}^{M}\sum_{j'=1}^{M} a_j a_{j'} \left(\frac{w_j}{a_j} - \frac{w_{j'}}{a_{j'}}\right)^2 d_{jj'} + \sum_{j=1}^{M} \frac{w_j^2}{a_j}\alpha_j,$$

vide J.N.K. Rao (1979) and Chaudhuri & Pal (2002)

Let $\qquad\qquad C_j = E_p(b_{\delta j}{}^2 I_{\delta j} - 1)$

and $\qquad\qquad C_{jj'} = E_p(b_{\delta j}I_{\delta j} - 1)(b_{\delta j'}I_{\delta j'} - 1), j \neq j'.$

We may then write

$$V_p(t_b) = \sum_{j=1}^{M} C_j w_j^2 + \sum_{j=1}^{M}\sum_{j'=1}^{M} C_{jj'} w_j w_{j'}.$$

It follows that, on finding $C_{\delta j}$, $C_{\delta j'}$ such that $E_p(C_{\delta j}I_{\delta j}) = C_j$, and $E_p(C_{sjj'}I_{sjj'}) = C_{jj'}$

$$v_1(t_b) = \sum C_{\delta j} w_j^2 \frac{I_{\delta j}}{\pi_j} + \sum\sum_{j \neq j'} C_{\delta jj'} w_j w_{j'} \frac{I_{\delta j}}{\pi_{jj'}}$$

and

$$v_2(t_b) = -\sum_{1 \leq j < j'}^{M}\sum^{M} a_j a_{j'} \left(\frac{w_j}{a_j} - \frac{w_{j'}}{a_{j'}}\right)^2 d_{sjj'}\frac{I_{sjj'}}{\pi_{jj'}} + \sum_{j=1}^{M} \frac{w_j^2}{a_j}\alpha_j \frac{I_{sj}}{\pi_j},$$

if $E_p(d_{sjj'}I_{sjj'}) = d_{jj'}$, provided $\pi_{jj'} > 0 \; \forall \; j \neq j'$ implying $\pi_j > 0 \; \forall \; j$ too, are two unbiased estimators $\hat{V}_p(t_b)$ for $V_p(t_b)$;

Ha'jek (1958), J.N.K. Rao (1979), Chaudhuri & Pal (2002), and Chaudhuri & Stenger (2005) are a few relevant references for the above topics, as also is Chaudhuri (2010).

We may illustrate a few specific forms of t_b, $V_p(t_b)$, $v_p(t_b)$ as follows:

$$t_{HT} = \sum_j \frac{w_j}{\pi_j} I_{sj}, \text{ due to Horvitz \& Thompson (1952)}$$

$$V_p(t_{HT}) = \sum_{j<j'}\sum(\pi_j\pi_{j'} - \pi_{jj'})\left(\frac{w_j}{\pi_j} - \frac{w_{j'}}{\pi_{j'}}\right)^2 + \sum_{j=1}^{M}\frac{\beta_j w_j^2}{\pi_j}$$

due to Chaudhuri & Pal (2002),

writing

$$\beta_j = 1 + \frac{1}{\pi_j} \sum_{\substack{(j'=1 \\ j'\neq j)}}^{N} \pi_{jj'} - \sum_{j=1}^{N} \pi_j,$$

or alternatively,

$$V_p(t_{HT}) = \sum_j w_j^2 (1 - \pi_j)/\pi_j + \sum_{j\neq j'} \sum C_{jj'} \frac{w_j w_{j'}}{\pi_j \pi_{j'}} (\pi_{jj'} - \pi_j \pi_{j'})$$

due to Horvitz & Thompson (1952) with unbiased estimators for $V_p(t_{HT})$ as

$$v_1(t_{HT}) = \sum_{j<j'} \sum (\pi_j \pi_{j'} - \pi_{jj'}) \left(\frac{w_j}{\pi_j} - \frac{w_{j'}}{\pi_{j'}} \right)^2 \frac{I_{sjj'}}{\pi_{jj'}} + \sum_j \frac{w_j^2}{\pi_j} \beta_j \frac{I_{sj}}{\pi_j}$$

and

$$v_2(t_{HT}) = \sum_j w_j^2 \left(\frac{1 - \pi_j}{\pi_j} \right) \frac{I_{sj}}{\pi_j} + \sum_{j\neq j'} \sum w_j w_{j'} \left(\frac{\pi_{jj'} - \pi_j \pi_{j'}}{\pi_j \pi_{j'}} \right) \frac{I_{sjj'}}{\pi_{jj'}},$$

provided $\pi_{jj'} > 0 \ \forall \ j \neq j' = 1, \ldots, M$ due respectively to Chaudhuri & Pal (2002) and Horvitz & Thompson (1952).

Of course, for designs admitting samples only with distinct units with a fixed multiplicity of each, β_j equals zero for every $j = 1, \ldots, M$ and Yates & Grundy (1953) gave their versions much earlier than Chaudhuri & Pal (2002) for the above $V_p(t_{HT})$ and $v_1(t_{HT})$ as $V_{YG}(t_{HT}) = V_p(t_{HT})|\beta_j = 0$ and $v_{YG}(t_{HT}) = v_1(t_{HT})|\beta_j = 0$.

To employ the estimator due to Horvitz & Thompson (1952), it is necessary to specify an appropriate sampling scheme rightly functional in a given circumstance. Here we may refer to discussions relevant in this case given by Brewer & Hanif (1983) and Chaudhari &Vos (1988).

One may otherwise employ Rao, Hartley & Cochran's (1962) strategy, i.e., a sampling scheme in combination with their method of estimation, as described in Chapter 2. Only, we have to replace y_i's for $i = 1, \ldots, N$ by w_j's for $j = 1, \ldots, M$ and other notations appropriately re-designed. Alternatively, any other strategy briefly narrated in Chapter 2 with requisite changes of y_i, $i = 1, \ldots, N$ into w_j, $j = 1, \ldots, M$ and associated concepts and notations may also be safely implemented if found otherwise appropriate.

3.2.2   Single Stage Sampling of 'Selection Units' with Constraint in Sample Size

For the sake of simplicity and definiteness, suppose from the M Selection Units a sample of $m(2 < m < M)$ Selection Units is drawn following Rao, Hartley & Cochran's scheme assuming normed size-measures $p_j \left(0 < p_j < 1, \sum_{1}^{M} p_j = 1 \right)$ are available at hand. With w_j's as used in network sampling, then, referring to Rao, Hartley & Cochran's strategy described in Chapter 2, $t_{RHC} = \sum_m w_i \dfrac{Q_i}{p_i}$ is Rao, Hartley & Cochran's unbiased estimator of $W = \sum_{j=1}^{M} w_j$ and hence of $Y = \sum_{1}^{N} y_i$. Let M_i be the number of units taken out of M Selection Units in the ith group $(i = 1, \ldots, m)$ formed by Simple Random Sampling Without Replacement method for applying the Rao, Hartley & Cochran's strategy. Let $\sum_m$ denote summing over the m groups, used in t_{RHC} above, $\sum_m \sum_m$ that over the distinct pairs out of these groups avoiding duplications, (w_i, Q_i, p_i) for the (w_j, Q_j, p_j)—values relevant for the ith group needed in Rao, Hartley & Cochran's scheme. Let

$$A = \frac{\sum_m M_i^2 - M}{M(M-1)}, B = \frac{\sum_m M_i^2 - M}{M^2 - \sum_m M_i^2}.$$

Then,

$$V(t_{RHC}) = A \left[\sum_{j=1}^{m} \frac{w_j^2}{p_j} - W^2 \right]$$

and

$$\hat{V}(t_{RHC}) = B\left[\sum_m Q_i \frac{w_i^2}{p_i^2} - t_{RHC}^2\right] = B\left[\sum_m \sum_m Q_i Q_{i'} \left(\frac{w_i}{p_i} - \frac{w_{i'}}{p_{i'}}\right)^2\right]$$

is Rao, Hartley & Cochran's unbiased estimator for $V(t_{RHC})$. Here of course $Q_i = p_{i_1} + \cdots + p_{iM_i}$, writing p_{iu} as the normed size-measure for the unit iu falling in the ith group, $u = 1, \ldots, M_i$.

Consequently,

$$\sum_m Q_i = \sum_{j=1}^M p_j = 1.$$

Suppose the network sample chosen corresponding to the selected Selection Units m in number is found to be too big and unmanageable to execute the survey of the sampled Observational Units. Then the network sampling may be 'constrained' to achieve reduction in the final size of the sample as shown below.

Recall that
$$w_j = \sum_{i \in A(j)} \frac{y_i}{m_i}, j = 1, \ldots, M$$

Let $C_j = |A(j)| \equiv$ the cardinality of $A(j)$, i.e., the number of Observational Units in $A(j), j = 1, \ldots, M$. Let $C(s) = \sum_{j \in s} C_j$ be too large. So, finding it difficult to survey the y_i values of all i's in $A(j)$ for $j \in s$, let us decide to choose independently from respective $A(j)$ sub-samples $B(j)$ of sizes d_j for $j \in s$ such that $2 \leq d_j \leq C_j, j \in s$ with $\sum_{j \in s} d_j = d(s)$ small enough compared to $C(s)$ so that y_i values for $i \in U_{j \in s} B(j)$ may be ascertained on a survey with reduced hassles. Let π_i denote the inclusion-probability with which an Observational Unit i in $A(j)$ is selected in $B(j)$ such that $0 < \pi_i < 1$ for $i \in A(j)$ and $\sum_{i \in A(j)} \pi_i = d_j, j = 1, \ldots, M$. For example, one may take $\pi_i = \dfrac{d_j}{C_j}$ if $i \in A(j)$ and from $A(j)$

a Simple Random Sample Without Replacement in d_j draws is taken as $B(j)$. Let $\pi_{ii'}(0 < \pi_{ii'} < 1)$ with $\sum\limits_{\substack{i'=1 \\ i' \neq i}}^{C_j} \pi_{ii'} = (d_j - 1)\pi_i$

and $\sum\limits_{i \neq i'}^{C_j}\sum\limits^{C_j} \pi_{ii'} = d_j(d_j - 1)$, $j = 1, \ldots, M$, denote the inclusion-probabilities of the distinct pairs i and i' of observational units being included in a sample from $A(j)$ in $B(j)$. For example, in case $B(j)$ is a Simple Random Sample Without Replacement from $A(j)$ of size d_j, then $\pi_{ii'} = \dfrac{d_j(d_j - 1)}{C_j(C_j - 1)}$ for $i \neq i' \in A(j)$ for $j = 1, \ldots, M$.

Now let us replace the estimator t_{RHC} for W by

$$e_{RHC} = \sum_m u_i \frac{Q_i}{P_i},$$

on taking

$$u_j = \sum_{i \in B(j)} \left(\frac{y_i}{m_i}\right)\left(\frac{1}{\pi_i}\right) = \sum_{i \in B(j)} \frac{a_i}{\pi_i}, \text{ writing } a_i = \frac{y_i}{m_i}.$$

Then, let E_1, E_2 be the operators for taking the expectations respectively by Rao, Hartley & Cochran's sampling scheme and generically in sampling $B(j)$ from $A(j)$ by a suitable scheme subject to the conditions noted above on π_i and $\pi_{ii'}$'s.

Then, $E_2(u_j) = \sum\limits_{i \in A(j)} \left(\dfrac{y_i}{m_i}\right) = w_j$ and writing $E = E_1 E_2$, one has $E_2(e_{RHC}) = t_{RHC}$, $E(e_{RHC}) = E_1(t_{RHC}) = W = Y$, i.e., e_{RHC} is an unbiased estimator for $W = Y$.

Also, writing V_1, V_2 respectively as operators for calculating variance with respect to RHC scheme and in respect generically of sub-sampling $B(j)$ from $A(j)$ with π_i, $\pi_{ii'}$'s as inclusion-probabilities of the first two orders as described above and $V = E_1 V_2 + V_1 E_2$ we may derive the following:

First note: $a_i = \dfrac{y_i}{m_i}$. Then,

$$u_j = \sum_{i \in B(j)} \left(\frac{a_i}{\pi_i} \right)$$

so that

$$E_2(u_j) = \sum_{i \in A(j)} a_i = w_j.$$

Also following Chaudhuri & Pal (2002) and Chaudhuri (2010), we get

$$V_2(u_j) = \sum_{\substack{i,i' \in A(j) \\ i \ i'}} (\pi_i \pi_{i'} - \pi_{ii'}) \left(\frac{a_i}{\pi_i} - \frac{a_{i'}}{\pi_{i'}} \right) + \sum_{i \in A(j)} \left(\frac{a_i}{\pi_i} \right) \alpha_i$$

writing

$$\alpha_i = 1 + \frac{1}{\pi_i} \sum_{\substack{(i'=1) \\ i' \neq i}}^{C_j} \pi_{ii'} - \sum_{1}^{C_j} \pi_i;$$

then

$$e_{RHC} = \sum_m u_i \frac{Q_i}{P_i}$$

and

$$E_2(e_{RHC}) = t_{RHC}, \quad V_2(e_{RHC}) = \sum_m V_2(u_i) \left(\frac{Q_i}{P_i} \right)^2.$$

So,

$$V(e_{RHC}) = E_1[V_2(e_{RHC})] + V_1[E_2(e_{RHC})]$$

Then

$$v_2(u_j) = \sum_{\substack{i,i' \in B(j) \\ i < i'}} \left(\frac{\pi_i \pi_{i'} - \pi_{ii'}}{\pi_{ii'}} \right) \left(\frac{a_i}{\pi_i} - \frac{a_{i'}}{\pi_{i'}} \right)^2 + \sum_{i \in B(j)} \left(\frac{a_i}{\pi_i} \right)^2 \alpha_i$$

clearly satisfies

$$E_2[v_2(u_j)] = V_2(u_j), \quad j = 1, \ldots, M.$$

Also, writing

$$A = \frac{\sum_m M_j^2 - M}{M(M-1)},$$

one may get

$$V_1[E_2(e_{RHC})] = \left[A \sum_{j=1}^{M} \frac{w_j^2}{p_j} - W^2 \right] = V_1(t_{RHC}) = E_1(t_{RHC}^2) - W^2$$

$$= E_1[E_2(e_{RHC}^2) - V_2(e_{RHC})] - W^2$$

leading to

$$E_1 V_2(e_{RHC}) = E_1 E_2(e_{RHC}^2) - A \sum_{j=1}^{M} \frac{w_j^2}{p_j} - (1 - A)W^2.$$

$$V_1(t_{RHC}) = A \left[\sum \frac{w_j^2}{p_j} - W^2 \right]$$

So, we get $V(e_{RHC}) = E_1 E_2[v_2(e_{RHC})] + V_1[t_{RHC}]$

on writing $v_2(e_{RHC})$ for an expression to be derived satisfying $E_2[v_2(e_{RHC})] = V_2(e_{RHC})$.

Again, on writing

$$v_1(t_{RHC}) = B \sum_{m} \sum_{m} Q_i Q_{i'} \left(\frac{w_i}{p_i} - \frac{w_{i'}}{p_{i'}} \right)^2$$

which satisfies

$$E_1 v_1(t_{RHC}) = V_1(t_{RHC})$$

on taking

$$B = \frac{\sum_{m} M_j^2 - M}{M^2 - \sum_{m} M_j^2},$$

we derive the following:

$$V(e_{RHC}) = E_1 E_2 \left[\sum_{m} \left(\frac{Q_j}{p_{j'}} \right)^2 \left\{ \sum_{\substack{i,i' \in B(j) \\ i<i'}} \left(\frac{\pi_i \pi_{i'} - \pi_{ii'}}{\pi_{ii'}} \right) \left(\frac{a_i}{\pi_i} - \frac{a_{i'}}{\pi_{i'}} \right)^2 \right. \right.$$

$$\left. \left. + \sum_{i \in B(j)} \left(\frac{a_i}{\pi_i} \right)^2 \alpha_i \right\} \right] + E_1[v_1(t_{RHC})]$$

on taking

$$v_2(e_{RHC}) = \sum_m \left(\frac{Q_j}{p_j}\right)^2 \left\{ \sum_{\substack{i,i' \in B(j) \\ (i<i')}} \sum \left(\frac{\pi_i \pi_{i'} - \pi_{ii'}}{\pi_{ii'}}\right) \right. $$

$$\left. \left(\frac{a_i}{\pi_i} - \frac{a_{i'}}{\pi_{i'}}\right)^2 + \sum_{i \in B(j)} \left(\frac{a_i}{\pi_i}\right)^2 \alpha_i \right\}$$

and

$$v_1(t_{RHC}) = B \sum_m \sum_m Q_i Q_{i'} \left(\frac{w_i}{p_i} - \frac{w_{i'}}{p_{i'}}\right)^2$$

Let

$$D = B \sum_m \sum_m Q_i Q_{i'} \left(\frac{u_i}{p_i} - \frac{u_{i'}}{p_{i'}}\right)^2$$

Then,

$$E_2(D) = B \sum_m \sum_m Q_i Q_{i'} \left[\left(\frac{V_2(u_i)}{p_i^2} + \frac{V_2(u_{i'})}{p_{i'}^2}\right) + \left(\frac{w_i}{p_i} - \frac{w_{i'}}{p_{i'}}\right)^2 \right]$$

So,

$$B \sum_m \sum_m Q_i Q_{i'} \left[\left(\frac{u_i}{p_i} - \frac{u_{i'}}{p_{i'}}\right)^2 - \left(\frac{v_2(u_i)}{p_i^2} + \frac{v_2(u_{i'})}{p_{i'}^2}\right) \right]$$

is an unbiased estimator for

$$B \sum_m \sum_m Q_i Q_{i'} \left(\frac{w_i}{p_i} - \frac{w_{i'}}{p_{i'}}\right)^2,$$

which is an unbiased estimator for $V_1(t_{RHC})$.

So, finally,

$$
v = \sum_m \left(\frac{Q_j}{P_j}\right)^2 \left[\sum_{\substack{i,i' \in B(j) \\ i < i'}} \left(\frac{\pi_i \pi_{i'} - \pi_{ii'}}{\pi_{ii'}}\right)\left(\frac{a_i}{\pi_i} - \frac{a_{i'}}{\pi_{i'}}\right)^2 + \sum_{i \in B(j)} \left(\frac{a_i}{\pi_i}\right)^2 \alpha_i \right]
$$

$$
+ B \sum_m \sum_m Q_i Q_{i'} \left[\left(\frac{u_i}{p_i} - \frac{u_{i'}}{p_{i'}}\right)^2 - \left(\frac{v_2(u_i)}{p_i^2} + \frac{v_2(u_{i'})}{p_{i'}^2}\right) \right]
$$

provides an unbiased estimator for $V(e_{RHC})$.
  Here, of course,

$$
u_j = \sum_{i \in B(j)} \left(\frac{y_i}{m_i}\right)\frac{1}{\pi_i} = \sum_{i \in B(j)} \left(\frac{a_i}{\pi_i}\right),
$$

$$
v_2(u_j) = \sum_{\substack{i,i' \in B(j) \\ i < i'}} \left[\left(\frac{\pi_i \pi_{i'} - \pi_{ii'}}{\pi_{ii'}}\right)\left(\frac{a_i}{\pi_i} - \frac{a_{i'}}{\pi_{i'}}\right)^2 \right] + \sum_{i \in B(j)} \left(\frac{a_i}{\pi_i}\right)^2 \alpha_i
$$

as noted earlier, and

$$
E_2[v_2(u_j)] = \sum_{\substack{i,i' \in A(j) \\ i \; i}} (\pi_i \pi_{i'} - \pi_{ii'})\left(\frac{a_i}{\pi_i} - \frac{a_{i'}}{\pi_{i'}}\right) + \sum_{i \in A(j)} \frac{a_i}{\pi_i} \alpha_i = V_2(u_j).
$$

  The formulae simplify as follows if $B(j)$ is taken as a simple Random Sample Without Replacement from $A(j)$ in d_j draws independently across $j = 1, \ldots, M$.

$$
e_{RHC} = \sum_m \left(\frac{Q_j}{P_j}\right) u_j, \quad u_j = \frac{C_j}{d_j} \sum_{i \in B(j)} a_i, \quad a_i = \frac{y_i}{m_i}
$$

Then,

$$
E_2(u_j) = \sum_{i \in A(j)} a_i = w_j
$$

$$
E_2(e_{RHC}) = \sum_m \frac{Q_j}{P_j} w_j = t_{RHC}
$$

$$
V_2(u_j) = C_j^2 \left(\frac{1}{d_j} - \frac{1}{C_j}\right)\frac{1}{(C_j - 1)} \sum_{i \in A(j)} \left(a_i - \frac{1}{C_j}\sum_{i \in A(j)} a_i\right)^2
$$

$$v_2(u_j) = C_j^2 \left(\frac{1}{d_j} - \frac{1}{C_j} \right) \frac{1}{(d_j - 1)} \sum_{i \in B(j)} \left(a_i - \frac{1}{d_j} \sum_{i \in B(j)} a_i \right)^2$$

for which $E_2 v_2(u_j) = V_2(u_j)$.

$$V_2(e_{RHC}) = \sum_m \left(\frac{Q_j}{p_j} \right)^2 V_2(u_j)$$

$$V(e_{RHC}) = E_1 V_2(e_{RHC}) + V_1(E_2(e_{RHC}))$$

$$= E_1 \left[\sum_m \left(\frac{Q_j}{p_j} \right)^2 V_2(u_j) \right] + V_1(t_{RHC})$$

$$= E_1 E_2 \left[\sum_m \left(\frac{Q_j}{p_j} \right)^2 v_2(u_j) \right] + A \sum_m \sum_m Q_i Q_{i'} \left(\frac{w_i}{p_i} - \frac{w_{i'}}{p_{i'}} \right)^2$$

Let $$D = B \sum_m \sum_m Q_i Q_{i'} \left(\frac{u_i}{p_i} - \frac{u_{i'}}{p_{i'}} \right)^2$$

Then,

$$E_2(D) = B \sum_m \sum_m Q_i Q_{i'} \left(\frac{w_i}{p_i} - \frac{w_{i'}}{p_{i'}} \right)^2 + \left[\frac{V_2(u_i)}{p_i^2} + \frac{V_2(u_{i'})}{p_{i'}^2} \right]$$

and

$$E_1 E_2(D) = V_1(t_{RHC}) + E_1 E_2 \left[B \sum_m \sum_m Q_i Q_{i'} \left\{ \frac{v_2(u_i)}{p_i^2} + \frac{v_2(u_{i'})}{p_{i'}^2} \right\} \right]$$

So,

$$v = \sum_m \left(\frac{Q_j}{p_j} \right)^2 v_2(u_j) + B \sum_m \sum_m Q_i Q_{i'} \left[\left(\frac{u_i}{p_i} - \frac{u_{i'}}{p_{i'}} \right)^2 - \left\{ \frac{v_2(u_i)}{p_i^2} + \frac{v_2(u_{i'})}{p_{i'}^2} \right\} \right]$$

provides an unbiased estimator for $V(e_{RHC})$.

Need for Adaptive Sampling

Need for Adaptive Sampling: Relevant Theory of Unbiased Estimation

4.1 INTRODUCTION

Adaptive sampling is a technique intended to enhance the information content of an initial sample chosen in a suitable way. It is often found worthy of attention and application in circumstances typified by some illustrations given below.

Thompson (1990, 1992), Thompson & Seber (1996), further probed into by Chaudhuri (2000) typically take up a practical situation as follows. In a geographically extensive country like India it is well-known that mineral oilfields are hard to come by. But because of an abiding need for fuel it is considered very important to explore locations where oil deposits may be sighted for commercial purposes. In such a search process it may be a good idea to circumscribe a large potential plot of land as a rectangular site. Splitting this up into a large number of smaller rectangular non-overlapping cells one may implement a suitably designed sample survey to carry out drilling a suitable number of such cells that may happen to fall in the sample actually chosen. Unfortunately only a negligible number of these selected plots may be found to contain enough oil contents, the rest being found empty or insignificant in content.

But the situation may not be so bad if instead of the empty cells some more in close neighbourhood of those with plenty of oil deposits were sampled and grilled. Such a compensatory sampling survey procedure may fruitfully be designed ensuring enhanced information contents. How can this be scientifically accomplished is elaborated in the subsequent sections.

4.2 ADAPTIVE SAMPLING SCHEME

Given a sampling unit say, i in a finite survey population $U = (1, \ldots, i, \ldots, N)$, it is important to define its neighbourhood. By the neighbourhood of a unit i we mean a set composed of itself plus other units in U which bear a reciprocal unalterable relationship so that for each of such other units when addressed, the initial unit will inalienably be referred to as bearing this reciprocal relationship. If an initial unit bears the characteristic of interest in an investigation, each of the units in its neighbourhood will be checked for its bearing the same characteristic. If it bears it, all its neighbouring units will also be checked for the same. This process is to continue until reaching the stage when none in the neighbourhood possesses the feature we are looking for. The entire set of such units checked for constitutes the cluster of units for the initial unit i. Any unit in the cluster not bearing the feature is called an "Edge Unit". The set of units in the cluster for i, omitting all the edge units in it, constitutes the "Network" for the initial unit i.

The estimation procedure in its elementary form is given in Section 1.1.

4.3 ADAPTIVE SAMPLING WITH SAMPLE-SIZE RESTRICTIONS

4.3.1 Summary

A practical difficulty in implementing an adaptive sampling design introduced by Thompson (1990) and vigorously studied by Thompson & Seber (1996) is the lack of an in-built check on the final size of the sample which is a random variable even though the initial sample-size may be suitably pre-assigned. Salehi & Seber (1997, 2002) among others cited by them, addressed and provided their solutions for this. We prescribe a simple alternative of Simple Random Sub-Sampling (SRSS) Without Replacement (WOR) independently from the "networks" of each initially sampled unit with a pre-assigned total for the sub-sample sizes. Modifications needed for estimation of the total of a variable of interest and an estimator for its measure of error are indicated for this revised adaptive sampling with a constrained sample-size.

Some Key Words: Adaptive sampling, Linear estimator of total, Restricted sample size.

4.3.2   **Introduction**

It was Thompson (1990) who introduced adaptive sampling while Thompson (1992) and Thompson & Seber (1996) later gave further details. Chaudhuri (2000) followed by Chaudhuri, Bose & Ghosh (2002) gave simple methods of estimating in an unbiased manner, the total of a variable of interest and the variance of an estimator of a population total illustrating its application in estimating numbers of rural industry-specific earners with concentrations in unknown and widely scattered locations.

Suppose there is a finite survey population of a known number N of units and our interest is to estimate the population total Y for a variable of interest y with y_i as its value for the ith unit of the population $U = (1, \ldots, i \ldots, N)$. Further, suppose y is zero-valued for many units of U though for many others its value is positive and quite high so that Y is supposedly a substantial number. For every unit a "neighbourhood" of units including itself is uniquely defined. For a given unit with a positive y-value, a cluster is formed by associating with it all the units in its neighbourhood and successively extending the association accommodating the units in the neighbourhoods of any unit with positive y-values and stopping only on encountering units with zero-values of y. The units with zero y values in the cluster are called the "edge-units" and the set of the remaining units in the cluster including the initial unit is called a "Network" of the latter. The edge-units are called "Singleton" networks. All the networks and the Singleton networks are 'non-overlapping' and together they exhaust the population.

Denoting by $A(i)$, the "Network" of a unit i in U and writing C_i as the cardinality of $A(i)$, it has been observed in the literature cited that the total T of $t_i = \dfrac{1}{C_i} \displaystyle\sum_{j \in A(i)} y_j$ over $i \in U$ equals Y. So, on choosing a sample s from U with a probability $p(s)$ the same formula involving $(y_i,\ i \in s)$ in estimating Y may be employed involving $(t_i,\ i \in s)$ in estimating T and hence Y itself. But in using $(t_i,\ i \in s)$, one has to ascertain y_j for every j in the networks $A(i)$ of every i in s. The set of units in the union of $A(i)$ over i in s is called an "Adaptive sample" reached through the initial sample s. Obviously, the size of the adaptive sample $A(s)$ is a random variable $v(s)$, say, with a value that may substantially exceed the effective size $n(s)$ of the initial sample s.

This is undoubtedly a potential repulsive feature of Adaptive sampling. This problem has of course been elegantly addressed with encouraging solutions by Salehi and Seber (1997, 2002), Brown (1994) and others. We offer a simple approach, permitting instead of a complete survey of all the units in the networks of the sampled units, that of only sub-samples of suitable sizes from them by Simple Random Sampling (SRS) Without Replacement (WOR), such that the sum of the sub-sample sizes is kept within a pre-assigned limit. Procedures of estimation of the total and that of a measure of error of the estimator of the total for such a proposed "size-constrained adaptive sampling scheme" are presented in Subsection 4.3.3 in a general set-up. In Subsection 4.3.4 is illustrated a couple of simple special cases.

4.3.3   Estimation of Total and of Measure of Error of Estimator

Following Rao (1979), Chaudhuri & Pal (2002) considered for $Y = \displaystyle\sum y_i$ a homogeneous linear estimator based on $(y_i,\ i \in s)$ for s chosen with probability $p(s)$, namely,

$$t = \sum y_i b_{si} I_{si} \tag{4.1}$$

with $I_{si} = 1$ if $i \in s$

$\qquad\quad = 0$ if $i \notin s$, Σ denoting sum over i in U

and b_{si} free of $\underline{Y} = (y_1, \ldots, y_i, \ldots, y_N)$. They also gave its Mean Square Error (MSE) as

$$M(t) = -\sum_{i<j}\sum d_{ij} w_i w_j \left(\frac{y_i}{w_i} - \frac{y_j}{w_j}\right)^2 + \sum \frac{y_i^2}{w_i}\alpha_i \qquad (4.2)$$

with w_i as any specified non-zero constants, and

$$d_{ij} = E_p(b_{si}I_{si} - 1)(b_{sj}I_{sj} - 1), \ \alpha_i = \sum_{j=1}^{N} d_{ij}w_j, E_p$$

denoting expectation with respect to the sampling design p specifying $p(s)$ above. Further, a generic form of an unbiased estimator for $M(t)$ is

$$m(t) = -\sum_{i<j}\sum d_{sij} I_{sij} w_i w_j \left(\frac{y_i}{w_i} - \frac{y_j}{w_j}\right)^2 + \sum C_{si}I_{si}\frac{y_i^2}{w_i}\alpha_i \qquad (4.3)$$

with d_{sij}, C_{si} free of $\underline{Y}$, $I_{sij} = I_{si}I_{sj}$ such that $E_p(d_{sij}I_{sij}) = d_{ij}$ and $E_p(C_{si}I_{si}) = 1$.

Following Chaudhuri (2000) and heeding the arguments in Subsection 4.3.2, it follows that based on an adaptive sample reached through s yielding $(t_i, i \in s)$, the counter-parts of Eqs (4.1)–(4.3) are derived simply by replacing y_i with t_i. But upto this level no constraint is imposed on the total effective size of the adaptive sample which is say

$$\sum_{i \in s} C_i = A(s).$$

From considerations of the resources at hand let us fix a number L and numbers d_i for $i \in s$ as the cardinalities of sub-sets $B(i)$ of $A(i)$, $i \in s$, such that

$$\sum_{i \in s} d_i \leq L$$

and choose $B(i)$'s as Simple Random Sub-Samples Without Replacement from $A(i)$'s 'independently' across $i \in s$.

Let
$$u_i = \frac{1}{d_i} \sum_{j \in B(i)} y_{j,i \in s} \qquad (4.4)$$

Writing E_R, V_R as expectation, variance operators with respect to this Simple Random Sampling Without Replacement of the subsets of $A(i)$'s, $i \in s$ and $E = E_p E_R$ as the overall expectation and $V = E_p V_R + V_p E_R$ as the overall variance operators, we may observe the following:

$$E_R(u_i) = t_i, V_R(u_i) = \left(\frac{1}{d_i} - \frac{1}{C_i}\right) \frac{1}{(C_i - 1)} \sum_{j \in A(i)} (y_j - t_i)^2$$

and

$$v_R(u_i) = \left(\frac{1}{d_i} - \frac{1}{C_i}\right) \frac{1}{(d_i - 1)} \sum_{j \in B(i)} (y_j - u_i)^2$$

satisfies $E_R v_R (u_i) = V_R (u_i)$, $i \in s$.

For this 'Constrained adaptive sampling' with the above size-restrictions, our proposed estimator for $Y = T$ is

$$e = \sum u_i b_{si} I_{si} \tag{4.5}$$

for which $E_R(e) = t'$, writing t' for t on replacing y_i by t_i in t and its Mean Square Error about $Y = T$ is

$$M(e) = E(e - T)^2 = E_p E_R [(e - E_R(e)) + (t' - T)]^2$$

$$= E_p E_R \left[\sum (u_i - t_i) b_{si} I_{si}\right]^2 + E_p (t' - T)^2$$

$$= E_p \left[\sum V_R(u_i) b_{si}^2 I_{si}\right] + M'(t')$$

$$= E_p E_R \left[\sum v_R(u_i) b_{si}^2 I_{si}\right] + E_p m'(t'), \tag{4.6}$$

writing $M'(t')$ for $M(t)$ with t_i substituting for y_i in the latter and likewise $m'(t')$ for $m(t)$.

Remark: The proposed procedure entails the cost of ascertaining C_i for $i \in s$ but restricts to ascertainment of y_j only for $j \in B(i)$, $i \in s$. It is understood that d_i's subject to $\sum_{i \in s} d_i \le L$ are so chosen that there is adequate reduction in the cost compared to the one needed to accomplish ascertainment of y_j for all j in $A(i)$ for $i \in s$.

Let

$$f_i = u_i^2 - v_R(u_i); \quad \text{then } E_R(f_i) = t_i^2 \tag{4.7}$$

$$a_{sij} = -\sum\sum_{i<j} d_{sij} I_{sij} w_i w_j \left(\frac{u_i}{w_i} - \frac{u_j}{w_j}\right)^2$$

$$b_{sij} = a_{sij} + \sum\sum_{i<j} d_{sij} I_{sij} w_i w_j \left[\frac{v_R(u_i)}{w_i^2} + \frac{v_R(u_j)}{w_j^2}\right]$$

Then,

$$E_R(a_{sij}) = -\sum\sum_{i<j} d_{sij} I_{sij} w_i w_j \left[\left(\frac{t_i}{w_i} - \frac{t_j}{w_j}\right)^2 + \left(\frac{V_R(u_i)}{w_i^2} + \frac{V_R(u_j)}{w_i^2}\right)\right]$$

and

$$E_R(b_{sij}) = -\sum\sum_{i<j} d_{sij} I_{sij} w_i w_j \left[\left(\frac{t_i}{w_i} - \frac{t_j}{w_j}\right)\right]^2.$$

So,

$$m(e) = \sum v_R(u_i) b_{si}^2 I_{si}$$

$$- \sum\sum_{i<j} d_{sij} I_{sij} w_i w_j \left[\left(\frac{u_i}{w_i} - \frac{u_j}{w_j}\right)^2 - \left(\frac{v_R(u_i)}{w_i^2} + \frac{v_R(u_j)}{w_j^2}\right)\right]$$

$$+ \sum \frac{f_i}{w_i} \alpha_i C_{si} I_{si} \tag{4.8}$$

is our proposed unbiased estimator for $M(e)$ because

$$Em(e) = E_p E_R \left[\sum v_R(u_i) b_{si}^2 I_{si} - \sum\sum_{i<j} d_{ij} w_i w_j \left(\frac{t_i}{w_i} - \frac{t_j}{w_j}\right)^2 + \sum \frac{t_i^2}{w_i} \alpha_i\right]$$

$$= M(e),$$

recalling Eqs (4.2)–(4.3).

4.3.4 Two Special Cases

Case 1. Horvitz–Thompson's Estimator
For the Horvitz and Thompson's (1952) unbiased estimator for Y, namely

$$t_H = \sum \frac{y_i}{\pi_i} I_{si} \tag{4.9}$$

$\pi_i = \sum_{s \ni i} p_s$, the inclusion-probability of i assumed positive, Chaudhuri (2000) and Chaudhuri & Pal (2002) gave the formulae for the variance as

$$V(t_H) = \sum_{i<j} \sum (\pi_i \pi_j - \pi_{ij}) \left(\frac{y_i}{\pi_i} - \frac{y_j}{\pi_j} \right)^2 + \sum \frac{y_i^2}{\pi_i} \beta_i, \tag{4.10}$$

$$\beta_i = 1 + \left(\frac{1}{\pi_i} \sum_{j \neq i} \pi_{ij} - \sum \pi_i \right), i \in U, \pi_{ij} = \sum_{s \ni i,j} p(s) \tag{4.11}$$

and for an unbiased estimator of $V(t_H)$ as

$$v(t_H) = \sum_{i<j} \sum (\pi_i \pi_j - \pi_{ij}) \left(\frac{y_i}{\pi_i} - \frac{y_j}{\pi_j} \right)^2 \frac{I_{sij}}{\pi_{ij}} + \sum \frac{y_i^2}{\pi_i} \beta_i \frac{I_{si}}{\pi_i}, \tag{4.12}$$

assuming $\pi_{ij} > 0 \ \forall i,j \in U$.

Substituting t_i throughout for y_i in Eqs (4.9), (4.10), (4.12), one may cover the case of Adaptive sampling with no constraint on the sample sizes, as is recommended implicitly by Chaudhuri's (2000) approach.

Imposing the sample-size constraints and implementing the procedure of network-wise sub-sampling and introducing the notations $E_R, V_R, B(i), u_i, d_i, f_i$ as in Subsection 4.3.3, in the present case our proposed 'unbiased' estimator for $T = Y$ based on the 'Adaptive sampling with constrained sample-size' is

$$e_H = \sum \frac{u_i}{\pi_i} I_{si} \tag{4.13}$$

Then, the variance of e_H is, writing $t'_H = \sum \frac{t_i}{\pi_i} I_{si}$

$$V(e_H) = E(e_H - T)^2 = E_p E_R \left[\sum \frac{(u_i - t_i)}{\pi_i} I_{si} + (t'_H - T) \right]^2$$

$$= E_p \left[\sum \frac{V_R(u_i)}{\pi_i^2} I_{si} \right] + V_p(t'_H) = \sum \frac{V_R(u_i)}{\pi_i} + V_p(t'_H) \tag{4.14}$$

with $V_p(t'_H)$ as $V(t_H)$ of Eq. (4.10) with y_i replaced by t_i, $i \in U$ in the latter.

Again, we shall write $v'(t'_H)$ for $v(t_H)$ with y_i in the latter of Eq. (4.12) replaced throughout by t_i, $i \in s$. Thus,

$$v'(t'_H) = \sum\sum_{i<j}(\pi_i\pi_j - \pi_{ij})\left(\frac{t_i}{\pi_i} - \frac{t_j}{\pi_j}\right)^2 \frac{I_{sij}}{\pi_{ij}} + \sum \frac{t_i^2}{\pi_i}\beta_i\frac{I_{si}}{\pi_i}.$$

Then, we propose for $V(e_H)$ the following unbiased estimator, namely

$$v(e_H) = \sum v_R(u_i)\frac{I_{si}}{\pi_i} + \sum\sum_{i<j}(\pi_i\pi_j - \pi_{ij})\left(\frac{u_i}{\pi_i} - \frac{u_j}{\pi_j}\right)^2\frac{I_{sij}}{\pi_{ij}}$$

$$+ \sum\frac{u_i^2}{\pi_i}\beta_i\frac{I_{si}}{\pi_i} \tag{4.15}$$

Its unbiasedness is supported by the following:

Theorem 1

$$Ev(e_H) = V(e_H).$$

Proof:

$$Ev(e_H) = E_pE_Rv(e_H)$$

$$= \sum V_R(u_i) + E_p$$

$$\left\{\sum\sum_{i<j}(\pi_i\pi_j - \pi_{ij})\left[\left(\frac{t_i}{\pi_i} - \frac{t_j}{\pi_j}\right)^2 + \left(\frac{V_R(u_i)}{\pi_i^2} + \frac{V_R(u_j)}{\pi_j^2}\right)\right]\frac{I_{sij}}{\pi_{ij}}\right.$$

$$\left. + \sum\frac{t_i^2}{\pi_i}\beta_i + \sum\frac{V_R(u_i)}{\pi_i}\beta_i\right\}$$

$$= \sum\frac{V_R(u_i)}{\pi_i} + V_p(t'_H)$$

Because

$$V_p(t'_H) = \sum\sum_{i<j}(\pi_i\pi_j - \pi_{ij})\left(\frac{t_i}{\pi_i} - \frac{t_j}{\pi_j}\right)^2 + \sum\frac{t_i^2}{\pi_i}\beta_i$$

and

$$\sum V_R(u_i) + \sum\sum_{i<j}(\pi_i\pi_j - \pi_{ij})\left(\frac{V_R(u_i)}{\pi_i^2} + \frac{V_R(u_j)}{\pi_j^2}\right) + \sum\frac{V_R(u_i)}{\pi_i}\beta_i$$

$$= \sum V_R(u_i)\left[1 + \frac{1}{\pi_i}\left(\sum\pi_i - \pi_i\right) - \frac{1}{\pi_i^2}\left(\sum\sum_{j\neq i}\pi_{ij}\right) + \frac{\beta_i}{\pi_i}\right]$$

$$= \sum\frac{V_R(u_i)}{\pi_i}\left[\sum\frac{1}{\pi_i}\sum_{j\neq i}\pi_{ij} + 1 + \frac{1}{\pi_i}\sum_{j\neq i}\pi_{ij} - \sum\pi_i\right], \text{ by Eq. (4.3)}$$

$$= \sum\frac{V_R(u_i)}{\pi_i}.$$

Case 2. Rao, Hartley and Cochran's Estimator. Rao, Hartley and Cochran's (1962) scheme involves selecting n distinct units from a population U, where the normed size-measures $p_i\left(0 < p_i < 1, \sum p_i = 1\right)$ are available. This consists in (1) randomly dividing U into n groups with N_i units taken in the ith group such that $N_i \geq 1$ for $i \in U$ and $\sum_n N_i = N$ and (2) 'independently' choosing from each group just one unit with a probability proportional to its normed size-measure from among the units falling in the respective groups, writing $\sum_n$ to denote summing over the n groups.

Denoting for simplicity by (p_i, y_i) the normed size-measure and the y-value for the unit chosen from the ith group formed as above, Rao, Hartley and Cochran's unbiased estimator for Y is

$$t_{RHC} = \sum_n y_i\frac{Q_i}{p_i} \tag{4.16}$$

writing Q_i as the sum of the normed size-measures of the N_i units falling in the ith group. We may note that $\sum_n Q_i = 1$.

Based on the Adaptive sample generated from the above Rao, Hartley and Cochran sample with no restriction on the size, the unbiased estimator for $T = Y$ is

$$e_R = e_{RHC} = \sum_n t_i \frac{Q_i}{p_i} \qquad (4.17)$$

Straightaway from Rao, Hartley and Cochran (1962), we have, on writing

$$A = \frac{\sum_n N_i^2 - N}{N(N-1)} \text{ and } B = \frac{\sum_n N_i^2 - N}{N^2 - \sum_n N_{i'}^2}$$

the variance of e_R as

$$V(e_{RHC}) = A\left(\sum \frac{t_i^2}{p_i} - T^2\right)$$

and Rao, Hartley and Cochran's unbiased estimator for $V(e_R)$ as

$$v_p(e_R) = B\sum_n Q_i\left(\frac{t_i}{p_i} - e_R\right)^2 = B\left(\sum_n t_i^2 \frac{Q_i}{p_i^2} - e_{RHC}^2\right) \qquad (4.18)$$

Again, applying our simple random sub-sampling without replacement procedure of Subsection 4.3.2 for the Adaptive sampling with constraints on the sample-size and using the obvious notations C_i, $A(i)$, $B(i)$, d_i, u_i, f_i, etc. applicable to this Rao, Hartley and Cochran's sample, our proposed unbiased estimator for $T = Y$ is

$$g = \sum_n u_i \frac{Q_i}{p_i} \qquad (4.19)$$

Let us recall and introduce the notation

$$f_i = u_i^2 - v_{RHC}(u_i)$$

and
$$v_{RHC}(g) = \sum_m \left(\frac{Q_i}{p_i}\right)^2 v_{RHC}(u_i).$$

Then, we have the

Theorem 2

$$\hat{v}(g) = v(g) + B\left[\sum_n \frac{Q_i}{p_i^2} f_i - \left(g^2 - \sum_n \left(\frac{Q_i}{p_i}\right)^2 v(u_i)\right)\right]$$

is an unbiased estimator for $V(g)$.

Proof: $V(g) = E_p V_R(g) + V_p E_R(g)$

$$= E_p\left[\sum_n \left(\frac{Q_i}{p_i}\right)^2 V_R(u_i)\right] + V_p(e_R).$$

Now, $V(g) = E(g^2) - e_R^2 = \sum_n \left(\frac{Q_i}{p_i}\right)^2 V(u_i).$

So, $E_R\left[g^2 - \sum_n \left(\frac{Q_i}{p_i}\right)^2 v_R(u_i)\right] = e_R^2$

$E_R(f_i) = E_R(u_i^2) - V_R(u_i) = t_i^2 .$

So, $E\hat{v}(g) = E_p\left[\sum_n \frac{Q_i^2}{p_i} V_R(u_i)\right] + BE_p\left[\sum_n \frac{Q_i}{p_i^2} t_i^2 - e_R^2\right]$

$$= E_p V_R(g) + E_p(v_p(e_R)) = E_p V_R(g) + V_p(e_R) = V(g).$$

A more elegant form of $\hat{v}(g)$ seems to be

$$\hat{v}(g) = \sum_n \left(\frac{Q_i}{p_i}\right)^2 v_R(u_i) + B\left[\sum_n \frac{Q_i}{p_i^2} f_i - (g^2) - \sum_n \left(\frac{Q_i}{p_i}\right)^2 v_R(u_i)\right]$$

$$= (1 + B)\sum_n \left(\frac{Q_i}{p_i}\right)^2 v_R(u_i) + B\left[\sum_n \frac{Q_i^2}{p_i^2} f_i - g^2\right] \qquad (4.20)$$

which is our final proposed estimator of $V(g)$.

CHAPTER

5

Adaptive and Network in Tandem with Constraints

Adaptive and Network Sampling in Tandem: Illustrating the Advantages

Abstract

Recently we encountered two practical survey sampling problems—one of estimating household expenses on health of the inmates treated at least for sometime last year as hospital in-patients and the other about living conditions of Child Labourers currently in a country. In the former, healthcare units and in the latter the Households and industrial/commercial establishments yielded easy pickups of the 'Selection Units' for sampling. To them are 'linked' feasible respectively the Households with inmates treated as in-patients and the Child Labourers traceable from the Selection Units as our "Observational Units". As the Observational Units in both cases of relevance are relatively scarce and scattered in unknown manners we propose application of well-known Network sampling in combination with Adaptive sampling technique as feasible prescriptions to ensure their adequate capture. Relevant estimation procedures are recounted.

5.1   INTRODUCTION

In a usual socio-economic survey of household expenses, employment, health, education, etc. it is enough to sample Households from various cross-sections in geographical entities by stratified multi-stage sampling with equal or varying probabilities. But it may not be an adequately informative and economically viable procedure if we deal with certain specific socio-economic issues. For example, we may like to study the living conditions of pedlars procuring and supplying merchandise from and to several fixed establishments. Or, we may intend to enumerate labourers working in small-scale firms in the rural areas. Or, our objective may be to estimate the average household expenses on the treatment of the inmates suffering as in-patients in healthcare units. In such a case, data from even numerous Households may not throw up any information that we require. Or, we may intend to study the living conditions of Child Labourers who may be thinly scattered in houses and in commercial/industrial establishments. In such situations more penetrative efforts seem necessary beyond mere house-to-house and/or unit-to-unit enquiries.

We recommend a combination of Network Sampling and Adaptive Sampling to cover such queries with provisions for keeping check on the growing sample sizes. In Chapter 2, a theory was presented covering an application to a given district with a limited area and population followed by an extension to a national canvas. Relevant references are Thompson (1992), Thompson & Seber (1996), Chaudhuri (2000), Chaudhuri, Bose & Ghosh (2004), Chaudhuri & Saha (2006), and Chaudhuri, Bose & Dihidar (2005) among others.

5.2   METHODS OF SAMPLE SELECTION AND ESTIMATION

If we base on the Indian situation, then we may find the cities and towns composed of urban blocks with approximately same population sizes of about 1000 each. With an empirically established common urban Household size, of roughly 4.5, the number of Households plus that of Establishments in each urban block may be roughly calculated. They may be stratified working out a handy number and from each stratum a small Simple Random Sample Without Replacement may be chosen, of course after having drawn a Simple Random Sample

Without Replacement of Urban Blocks of suitable sizes from the cities and separately from the combined group of all the towns in the district.

The villagers on the other hand may vary considerably in their population sizes roughly ascertainable from recent population censuses. So, using population sizes as size-measures, villages may be selected by Rao-Hartley-Cochran (1962) scheme in adequate numbers. The urban blocks are the First Stage Units and so are the villages. Indian villages, for example, have an average household size of 5.5. So, the number of Households and establishments in each village may be roughly worked out as the Second Stage Units as in the urban blocks. The rural Second Stage Units may also be stratified within each village in suitable numbers. Each stratum size being roughly ascertained, Simple Random Sample Without Replacements may be taken in suitable numbers of the Second Stage Units strata-wise. If, as in the case of surveying pedlars, surveying establishments alone, leaving alone the residential Households seems more apt, that course may be taken, after listing the establishments stratum-wise.

The following notations will be used in estimation:

i : Observational Unit; e.g. household, if the survey pertains to Household expenses on the treatment of Household members if then were in-patients in healthcare centres, or a Child Labourer, if the survey pertains to Child Labourer, or a pedlar.

j : A First Stage Unit, viz. a village or a block or a ward in a town or a city.

k : A Second Stage Unit, which is a household or an establishment.

A kth Second Stage Unit in a jth First Stage Unit will be referred to as a 'Selection Unit' because one may employ a probability-based procedure for selection of it out of all such conceivable units.

m_i : Number of Selection Units 'linked' to ith Observational Unit.

A_{jk} : Set of Observational Units to which is linked the kth Second Stage Unit of the jth First Stage Unit (strata-wise if strata of Second Stage Units are formed within First Stage Units to simplify verification of links among Observational Units and Selection Units and sample-Selection Hassles).

$$w_{jk} = \sum_{i \in A_{jk}} \frac{y_i}{m_i}$$

y_i = Value of a real variable y defined on the ith observational unit.

Result

Letting $\sum_{i}$ as the sum over all the Observational Units in a district of interest and $\sum_{j}\sum_{k}$ as the sum of all the k's within the jth First Stage Unit followed by that of all the j's in the district above, it follows that $\sum_{j}\sum_{k} w_{jk} = \sum_{i} y_i$.

5.2.1 An Observation

Our immediate objective to estimate $\sum_{i} y_i$ will be achieved on presenting a method to accomplish estimation of $\sum_{j}\sum_{k} w_{jk}$. Further notations:

T_{jk} = "Network" of kth Second Stage Unit of jth First Stage Unit. By "Network" we mean the following: every Selection Unit is supposed to have a well-defined "Neighbourhood" consisting of itself plus a limited number of other Selection Units in the population or in any specified part thereof, each of which is capable of giving the whereabouts of none or one or more Observational Units, with the condition that when similarly approached, each such Selection Unit may reciprocally recognize and refer to each other. If an initial Selection Unit divulges information on at least one Observational Unit, each in its neighbourhood has to be interrogated and if the latter yields the whereabouts of at least one Observational Unit, it will be similarly interrogated. The process is to be continued until none in neighbourhood has anymore facts to divulge about any Observational Unit. A Selection Unit incapable of giving the whereabouts of at least one Observational Unit is called an "Edge Unit" and the Selection Units reached starting with the initial Selection Unit are referred to as a "Cluster" of the initial Selection Unit. Leaving aside all the edge units of the cluster, the set of the remaining units is referred to as the "Network" of the initial Selection Unit. Recognizing every edge unit as a "Singleton Network", the 'Networks' are 'mutually disjointed' and they together exhaust the population of all the Selection Units in it i.e. the district.

$N_{(j,k)}$ = The set of Observational Units linked to T_{jk}.

C_{jk} = The number of Observational Units in $N_{(j,k)}$, i.e., the cardinality of $N_{(j,k)}$.

$$t_{jk} = \frac{1}{C_{jk}} \sum_u \sum_{u' \in N(j,k)} w_{uu'}.$$

An observation

$$\sum_j \sum_k t_{jk} = \sum_u \sum_{u' \in N(j,k)} w_{uu'}.$$

This is because the union of all the networks coincides with the union of Second Stage Units in all the First Stage Units cumulated over all the First Stage Units in the district.

$B(j, k)$ = A Simple Random Sample Without Replacement of d_{jk} ($2 \leq d_{jk} \leq C_{jk}$) Observational Units from C_{jk} Observational Units in $N_{(j,k)}$.

$D = \sum_J \sum_K d_{jk}$ = A pre-assigned number which is the maximum number of Observational Units permitted to be surveyed with the limited resources and budgetary constraints.

$$e_{jk} = \frac{1}{d_{jk}} \sum_{u,u' \in B(j,k)} w_{uu'}$$

s_{hj} = Sample of Second Stage Units from hth stratum in the jth First Stage Unit taken as Simple Random Sample Without Replacement, s = A sample of First Stage Units.

L_{hj} = Number of Second Stage Units in the hth stratum in jth First Stage Unit.

l_{hj} = Number of Second Stage Units chosen from L_{hj} Second Stage Units in the hth stratum in jth First Stage Unit.

H = Total number of strata in First Stage Unit.

H_1 = Number of strata for which $l_{hj} \geq 2$.

H_2 = Number of strata for which $l_{hj} = 1$.

$$L'_j = \sum_{h=1}^{H_2} L_{hj}.$$

$$l'_j = \sum_{h=1}^{H_2} l_{hj} = H_2$$

M = Number of First Stage Units in the district.

m = Number of First Stage Units sampled by Simple Random Sampling Without Replacement.

N = Total, unknown but finite number of Observational Units to which are linked the entire collection of Selection Units for k varying over the strata of Second Stage Units in the respective First Stage Units over all the j's from 1 through M covering the entire district.

$Y = \sum_{i=1}^{N} y_i$, the principal parameter we need to estimate,

$X = \sum_{i=1}^{N} x_i$, with x_i as the value of another variable x, on the ith observational unit and $\overline{Y} = \dfrac{Y}{N}$ and $R = \dfrac{Y}{X}$ are other mean and ratio parameters which may also be needed to estimate.

First, assuming a simple Random Sample of m First Stage Units is taken without Replacement from M First Stage Units, let us consider the following:

$$\alpha_{hj} = \frac{L_{hj}}{l_{hj}} \sum_{k\in sjh} t_{jk}, \beta_j = \frac{L_{hj}}{l_{hj}} \sum_{k\in sjh} e_{jk}$$

$$\alpha_j = \sum_{h=1}^{H} \alpha_{hj}, \beta_j = \sum_{h=1}^{H} \beta_{hj}$$

$$\hat{e} = \frac{M}{m} \sum_{j=1}^{m} \left[\sum_{h=1}^{H} \frac{L_{hj}}{l_{hj}} \sum_{k\in s_{jh}} e_{jk} \right] = \frac{M}{m} \sum_{j=1}^{m} \beta_j \quad \text{is our proposed un-}$$

biased estimator for Y.

$$\hat{f} = \frac{M}{m} \sum_{j=1}^{m} \left[\sum_{h=1}^{H} \frac{L_{hj}}{l_{hj}} \sum_{k\in s_{jh}} t_{jk} \right] = \frac{M}{m} \sum_{j=1}^{m} \alpha_j$$

$$\hat{W} = \frac{M}{m} \sum_{j=1}^{m} \sum_{h=1}^{H} \frac{L_{hj}}{l_{hj}} \sum_{k\in s_{jh}} w_{jk} \tag{i}$$

$$v(\hat{W}) = \hat{V}(\hat{W}) = M^2 \left(\frac{1}{m} - \frac{1}{M} \right) \sum_{j=1}^{m} \psi_j \tag{ii}$$

$$\psi_j = \sum_{h=1}^{H_1} L_{hj}^2 \left(\frac{1}{l_{hj}} - \frac{1}{L_{hj}} \right) \sum_{k=1}^{L_{hj}} \left(w_{jk} - \frac{\sum_{k=1}^{l_{hj}} w_{jk}}{l_{hj}} \right)^2$$

$$+ (L_j')^2 \left(\frac{1}{l_j'} - \frac{1}{L_j'} \right) \frac{1}{(H_2 - 1)} \sum_{1}^{H_2} \left(w_{jk} - \frac{\sum_{k=1}^{u_2} w_{jk}}{H_2} \right)^2$$

$$v(\hat{f}) = \hat{V}(\hat{f}) = v(\hat{W}) \big|_{wjk = t_{jk}}$$

$$v(s) = \frac{M}{m} \sum_{j=1}^{m} \sum_{h=1}^{H} \left(\frac{L_{hj}}{l_{hj}} \right)^2 \sum_{k \in s_{jh}} v_{jk}$$

$$v_{jk} = \left(\frac{1}{d_{jk}} - \frac{1}{C_{jk}} \right) \frac{1}{(d_{jk} - 1)} \sum_u \sum_{u' \in B(j,k)} (w_{uu'} - e_{jk})^2$$

$$v(\hat{e}) = \hat{V}(\hat{e}) = v(\hat{f}) \big|_{t_{jk} = e_{jk}} + v(s). \tag{iii}$$

This is an unbiased estimator for the variance of our proposed unbiased estimator $\hat{e}$ for Y.

$$\hat{g} = \hat{e} \big|_{y_i = x_i, i = 1, \ldots, N}$$

$$\hat{R} = \frac{\hat{e}}{\hat{g}}, \text{ a ratio estimator for } R = \frac{y}{x}.$$

$m(\hat{R})$ = an estimate for the Mean Square Error of $\hat{R}$ about R is proposed to be taken as

$$m(\hat{R}) = \frac{1}{(\hat{g})^2} \, v(\hat{e}) \,|_{y_i = y_i - \hat{R}_{x_i \forall i}}.$$

Next, suppose the First Stage Units which are the villages with known size-measures $z_j (j = 1, \ldots, M)$, taken as total village populations or other known positive integer-values, are selected by Rao, Hartley & Cochran's (1962) scheme. For these, the following notations are additionally needed.

M_g = Number of First Stage Units taken in the gth group ($g = 1, \ldots, m$) of First Stage Units chosen by Simple Random Sampling Without Replacement method in applying the Rao, Hartley & Cochran's scheme.

$$\sum_m = \text{Sum over the } m \text{ groups}$$

$$\sum_m M_g = M$$

M_g's are chosen optimally as

$$M_g = \left[\frac{M}{m}\right] \text{ for } g = 1, \ldots, m_1, = \left[\frac{M}{m}\right] + 1 \text{ for } g = m_1 + 1, \ldots, m$$

such that $\sum_m M_g = M.$

C_g = Sum of z_j' 's for j's in the gth group.

$$T = \sum_m C_g$$

$$p_j = \frac{z_j}{T'}$$

$$Q_j = \frac{C_g}{T} \text{ if } j\text{th First Stage Unit falls in the } g\text{th group.}$$

$$\frac{Q_j}{p_j} = \frac{C_g}{z_j}$$

$$B = \frac{\dfrac{\sum_m M_g^2 - M}{M}}{M^2 - \sum_m M_g^2}$$

$$\sum_n = \text{Sum over a 'Sub-sample' of } n \ (< m) \text{ groups chosen by}$$

Simple Random Sampling Without Replacement scheme out of all the m groups formed in Rao, Hartley & Cochran sampling.

$$\hat{t} = \frac{m}{n}\sum_n \frac{Q_j}{p_j}\beta_j \equiv \text{the proposed unbiased estimator for } Y.$$

$$A_j = \left(\frac{Q_j}{p_j}\right)^2 \psi_j$$

$$(\hat{y}_j^2) = \beta_j^2 - \psi_j$$

$$(\hat{t}^2) = (\hat{t})^2 - \left(\frac{m}{n}\right)^2 \sum_n A_j$$

Result

$$v(\hat{t}) = \hat{V}(\hat{t})$$

$$= \left(\frac{m}{n}\right)^2 \sum_n A_j + (1+B)\left[\frac{m^2\left(\frac{1}{n} - \frac{1}{m}\right)}{(n-1)}\left\{\sum_m \left(\frac{Q_j}{p_j}\right)^2 (\hat{y}_j^2)\right.\right.$$

$$\left.\left. - \left(\frac{n}{m^2}\right)(\hat{t})^2 - \left(\frac{m}{n}\right)^2 \sum_n A_j\right\}\right]$$

$$+ B\left[\left(\frac{m}{n}\right)\sum_n Q_j \frac{(\hat{y}_j^2)}{p_j^2} - (\hat{t}^2)\right]$$

is an unbiased estimator of the variance of $\hat{t}$.

This is the prime result in this discussion. The proof for this is easily derived with certain easy algebra implemented on the results mostly available in Chaudhuri & Saha (2006).

For any estimator $\hat{\theta}$ for a parameter θ equal to $Y, \overline{Y}, R$, say, $CV = 100\frac{\sqrt{m(\hat{\theta})}}{\hat{\theta}}$, where $m(\hat{\theta})$ is an estimator for the Mean Square Error equaling the variance V, if $\hat{\theta}$ is unbiased for θ.

Next, in order to produce a national estimate along with its standard error and coefficient of variation we introduce the following further notations.

So long we considered the Selection Units spread in a given district and the Observation Units also linked to the Selection Units in that particular district. The Selection Units were the Second Stage Units each contained strata-wise within a First Stage Unit for the respective First Stage Units in the district.

The total $Y = \sum_{1}^{N} y_i$ matches $W = \sum_j \sum_k w_{jk}$ which in its turn matches $T = \sum_j \sum_k t_{jk}$, the former on the strength of the 'links' among Observation Units and Selection Units and the latter because of the uniqueness of the networks once the boundaries are well-defined and so are the networks which are non-overlapping and co-extensive with the union over the Second Stage Units strata- and First stage unit-wise and over all the latter in the district.

We have shown how the total of a variable y for an Observation Unit over all the Observation Units in a district is to be estimated once we may estimate W or T and similarly for the mean of y's and the ratio of the total of y to that of x. Now, just extending the w_{jk}'s to all the districts in a country it is a simple matter to estimate the total, mean and ratio relating to the Observation Units in the country. We proceed as follows:

Notations to be added:

P = Number of provinces (like states and union territories in India) in the country.

D_u = Number of districts in the uth province $(u = 1, \ldots, P)$.

W_d = The total of the values of w_{jk} in the dth district $(d = 1, \ldots, D_u)$.

We have already seen how for a typical dth district W_d is to be unbiasedly estimated and how also the corresponding mean and ratio of two totals. We shall be through if we may show how the country total is to be unbiasedly estimated, providing in addition the variance or Mean Square Error estimators.

$$C = \sum_{u=1}^{P} \sum_{d=1}^{D_u} W_d \text{ is the country total.}$$

$a =$ The number of provinces sampled by employing the Rao, Hartley & Cochran's scheme using normed size-measures generically taken as p_j's and Q_s's as the sums of the p-values within the respective groups formed in applying the Rao, Hartley & Cochran's scheme on optimally taking P_j provinces in the jth group, $j = 1,\ldots,a$.

$\displaystyle\sum_a =$ Sum over the 'a' groups.

$$A = \frac{\displaystyle\sum_a P_j^2 - P}{P^2 - \displaystyle\sum_a P_j^2}; \quad A_u = \frac{\displaystyle\sum_{bu} P_{du}^2 - D_u}{D_u^2 - \displaystyle\sum_{bu} P_{du}^2}.$$

$b_u =$ Number of districts sampled from the D_u districts in the uth province.

$P_{du} =$ Number of districts falling in the d_uth group formed optimally in applying the Rao, Hartley & Cochran's scheme to draw a sample of b_u districts from D_uth group formed optimally in applying the Rao, Hartley & Cochran's scheme to draw a sample of b_u districts from D_u districts.

Q_{du}'s are sums of the p_{du}'s within the groups formed in applying the Rao, Hartley & Cochran's scheme.

$$\hat{C} = \sum_a \frac{Q_p}{P_p} \sum_{bu} \left(\frac{Q_{du}}{P_{du}}\right) \hat{W}_{du} \quad \text{is an unbiased estimator for}$$

C, $\hat{W}_{du}$ is an unbiased estimator for W_{du}.

$$v = \hat{V}(\hat{C}) = A\left[\sum_a \left(\frac{Q_p}{P_p}\right)^2 - (\hat{C})^2\right] + \sum_a \left(\frac{Q_{du}}{P_{du}}\right) v_u \quad \text{is an unbi-}$$

ased estimator of the variance of $\hat{C}$.

$$\hat{W}_u = \sum_{bu} \left(\frac{Q_{du}}{P_{du}}\right) \hat{W}_{du}.$$

$$v_u = \hat{V}(\hat{W}_u) = A_u\left[\sum_{bu} Q_{du}\left(\frac{\hat{W}_{du}}{P_{du}}\right)^2 - (\hat{W}_u)^2\right] + \sum_{bu}\left(\frac{Q_{du}}{P_{du}}\right) v_{du}$$

where $v_{du} = \hat{V}(\hat{w}_{du})$, is an unbiased estimator of the variance of $\hat{W}_u$.

$\hat{W}_u$ above is the same unbiased estimator of the total for the uth district as obtained by the formula for $\hat{W}$ in (i) and v_{du} is $v(\hat{w})$ as given in (ii).

Replacing $v(\hat{W})$ in (ii) by $v(\hat{e})$ in (iii) and W_{du} by $\hat{C}$ and using them in (iii) we obtain our final estimator in (iv) and (v) below which are

$$C^* = \hat{C}\,|_{\hat{W}_{du} = \hat{e}} \tag{iv}$$

$$v_{c^*} = \hat{V}(C^*) = \hat{V}(\hat{C})\,|_{\hat{W}_{du} = \hat{e}}. \tag{v}$$

CHAPTER

6

Applications and Case Studies

6.1   INTRODUCTION

Maternal mortality, child labour, drug abuse are some of the important problem areas demanding attention of the social scientists globally. Some experiences of the author's involvement are narrated illustrating use of emerging statistical methods.

6.2   HOW TO CONSTRAIN EXPLOSIVE SIZE OF A NETWORK SAMPLE? THEORY AND APPLICATION WITH ILLUSTRATED LIVE DATA

Taking the cue from some of our predecessors named in Section 6.2.1, we recognize Network Sampling as one in which there are two kinds of units called (1) 'Selection Units' with known (a) total sizes and (b) frames facilitating sample selection in sophisticated manners and (2) 'Observation Units' with these elements (a) and (b) unknown. In order to suitably estimate the total of real values of a variable over the unknown total of all such Observation Units in a context we claim, Network Sampling is appropriate. Theories of unbiased estimation along with estimated measures of error in estimation especially on imposing requisite constraints on increasing sample size are developed admittedly on sedate utilization of results from earlier works. A complete sample-survey-based exercise is illustrated below to demonstrate the efficacy of our approach.

6.2.1 Introduction

For a given district covering a wide area and a large population in an Indian State we consider a practical problem of estimating the total household expenses in providing institutional treatments to their inmates found to have suffered within a specified year-long period from one or more of three locally important diseases (i) related to heart, (ii) cancer and (iii) gallbladder requiring admission into hospitals/nursing homes/clinics as in-patients.

Executing the standard sample survey method of "stratified, cluster sampling in one or more stages with equal or unequal probabilities with or without replacement" this task of course may be accomplished. But it is not hard to imagine that many selected households visited may not have any member of the above type at all. So, anticipating inadequate information content in such a standard sample survey, we recommend undertaking "Network Sampling", instead, adapting a prescription by Thompson (1990, 1992), Thompson & Seber (1996) we adopt their formulation as follows in brief.

Let there exist an unknown number N of "units" labeled i from 1 through N bearing consecutive unknown real values y_i, $i \in U$, writing $U = (1, ..., N)$. A frame for this U is supposedly unavailable but a serviceable estimator is needed for the "Population Total"

$$Y = \sum_{1}^{N} y_i .$$

These units are called "Observation Units". On the other hand, let there exist a known number of M units with identifiable labels j from 1 through M with a frame suitable for selection of a sample s, say, of these units called "Selection Units", on adopting any suitable selection scheme. We suppose there exist certain "links" among the Observation Units and the Selection Units which may be sensibly exploited in implementing the estimation mentioned above.

Let us define the symbols:

(A) A_j is the set of Observation Units linked to a jth Selection Unit.

(B) m_i is the number of Selection Units linked to an Observation Unit labeled i.

To begin with, it is not easy to ascertain A_j, $j = 1, \ldots, M$ and m_i, $i = 1, \ldots, N$. But if a sample s of Selection Units is selected, say, with a probability $p(s)$, it may be possible, exploiting the "Selection Unit" versus Observation Unit links to recognize the mutual reciprocity and thus identify all the Observation Units contained in A_j for j in s and also ascertain the corresponding values of m_i Observation Units for i in A_j for j in s. This is "Network Sampling", the set of Observation Units linked to the respective Selection Units occurring in sample s, providing a 'Network'. In Subsection 6.2.2 we present, utilizing the existing literature, a solution for estimating Y.

Recognizing that even for a sample s with a moderate size, the size of a network sample may be prohibitively large, in Subsection 6.2.3 we present our novel work on "Constrained Sample Size" network sampling needed to delimit the explosive size of a network sample and offer results on corresponding adjustments in estimation. In Subsection 6.2.4 we present a relevant case study mentioned earlier. Subsection 6.2.5 gives some concluding comments. Certain relevant references are Chaudhuri (2000), Chaudhuri & Stenger (2005, pp 314-316) and Chaudhuri (2010, pp 114–116 and 366–377).

6.2.2 Unbiased Estimation of Observation Unit Total in Network Sampling

Following Chaudhuri (2000) and Chaudhuri & Stenger (2005), let us define

$$w_j = \sum_{j \in A_j} \frac{y_i}{m_i}, \; j = 1, \ldots, M.$$

Then

$$W = \sum_{j=1}^{M} w_j \text{ equals } Y = \sum_{i=1}^{N} y_i$$

because

$$W = \sum_{j=1}^{M} w_j = \sum_{j=1}^{M} \sum_{j \in A_j} \frac{y_i}{m_i} = \sum_{i=1}^{N} \left(\frac{y_i}{m_i} \right) \sum_{\left(j/A_j \ni i \right)} 1 = \sum_{i=1}^{N} y_i = Y$$

because

$$\sum_{j/A_j \ni i} 1 \text{ equals } m_i,$$

by definition, for $i \in U$.

A Remark

So, in order to estimate Y it is enough to estimate W. Since the Selection Units collectively have a frame, a probability sample of Selection Units may be suitably selected and an unbiased estimator for W is naturally derivable. For instance, we shall show how Rao, Hartley & Cochran's (1962) strategy that is selection-cum-estimation method may work.

Rao, Hartley & Cochran's selection scheme

Let $p_j \left(0 < p_j < 1, \sum_{j=1}^{M} p_j = 1 \right)$ be the numbers available for the respective Selection Units as their normed size measures. In order to choose a sample s of m ($2 \leq m < M$) Selection Units from the M Selection Units, m disjoint groups need to be formed taking Simple Random Samples (SRS) Without Replacement (WOR) of M_1 Selection Units, then M_2 Selection Units out of $M - M_1$ Selection Units and so on and finally M_m Selection Units out of $M - \cdots - M_{m-1}$ Selection Units. Here Rao, Hartley & Cochran have recommended the choice

$$M_u = \left[\frac{M}{m} \right] \quad \text{for} \quad u = 1, \ldots, k$$

and

$$= \left[\frac{M}{m} \right] + 1 \quad \text{for} \quad u = k + 1, \ldots, m$$

on "uniquely" taking k such that for these M_u's $\sum_m M_u$ equals M. Here $\sum_m$ denotes summing over all the m groups formed. Writing $p_{u1}, p_{u2}, \ldots, p_{uM_u}$ as the normed size measures for the M_u Selection Units falling in the uth random group, let $Q_u = \sum_{k=1}^{M_u} p_{uk}$.

Then, let $t = \sum_m \dfrac{Q_u}{p_u} w_u$, writing w_u, p_u for the values of w_j, p_j for the Selection Unit falling in the uth group. Then, Rao, Hartley & Cochran have shown that $(i)t$ is an unbiased estimator for W.

(ii) The variance of t is

$$V(t) = A \sum_{j<}^{M} \sum_{j=1}^{M} p_j p_{j'} \left(\frac{w_j}{p_j} - \frac{w_{j'}}{p_{j'}} \right)^2$$

writing $\sum_{j<}^{M} \sum_{j=1}^{M}$ as the summation over all the $\binom{M}{2}$ non-duplicating pairs of units and $A = \dfrac{\sum_m M_u^2 - M}{M(M-1)}$.

(iii) An unbiased estimator for $V(t)$, writing $B = \dfrac{\sum_m M_u^2 - M}{M^2 - \sum_m M_u^2}$,

is $v(t) = B \sum_m \sum_m Q_u Q_{u'} \left(\dfrac{w_u}{p_u} - \dfrac{w_{u'}}{p_{u'}} \right)^2$. $\sum_m \sum_m$ is the sum of all

the $\binom{m}{2}$ pairs of disjoint groups formed by RHC's scheme discarding repititions.

It may be emphasized that verifiably $A > 0, B > 0$.

In passing we may add that if s is taken as a Simple Random Sample Without Replacement of m Selection Units out of M, then

$$V(t') = M^2 \left(\frac{1}{m} - \frac{1}{M} \right) \frac{1}{M-1} \sum_{j=1}^{M} \left(w_j - \frac{\sum_{j=1}^{M} w_j}{M} \right)^2$$

and

$$
v(t') = M^2 \left(\frac{1}{m} - \frac{1}{M} \right) \frac{1}{m-1} \sum_{j \in s} \left(w_j - \frac{\sum\limits_{j \in s} w_j}{m} \right)^2
$$

is an unbiased estimator for $V(t')$. Thus our problem of unbiased estimation of the Observation Units total $Y = \sum\limits_{1}^{N} y_i$ is solved. Here we write t' for t when Simple Random Sampling Without Replacement is employed.

It is well known (c.f. Cochran (1977)) that accuracy in estimation may be judged from the magnitudes of Coefficients of Variation of t and t' estimated as

$$
cv(t) = 100 \frac{+\sqrt{v(t)}}{t}
$$

and

$$
cv(t') = 100 \frac{+\sqrt{v(t')}}{t'} .
$$

A thumb rule for judgment is:

If $cv \le 10\%$, the estimator is excellent, if $10\% < cv \le 20\%$, the estimator is good enough; if $20\% < cv \le 30\%$, the estimator is still acceptable; and if $cv > 30\%$, the estimator may be discarded.

This way of judging is common in sample surveys because a best strategy never exists and one strategy can hardly be compared against another "uniformly for all possible variate values" unless one starts with an exceptional strategy which is inadmissible (cf. Cassel, Sarndal & Wretman (1977)).

6.2.3   Adjustment for "Constrained Sample Size" Network Sampling in Estimation

In the context of Adaptive cluster sampling of Thompson (1990, 1992) and Thompson & Seber (1996) as extended by Chaudhuri (2000) to cover general unequal probability sampling, Chaudhuri, Bose & Dihidar (2005) noted how even with a modest initial sample-size the terminating adaptive sample may have an explosive size causing an immense execution problem and recommended remedies through their "Sample-size-Restrictive Adaptive Sampling" presenting empirical applications. Chaudhuri (2003) presented algebra to accommodate deliberately implemented under-cover of an initial sample to economize time and money. Chaudhuri (2010) gave a case study in the context of a Child Labour Survey undertaken by him commissioned by International Labour Organization of the United Nation applying network sampling and adaptive sampling in tandem. In that connection also, constraining the sample size came as an important issue requiring his attention. In what follows we briefly present our thought in constraining the sample-size for network sampling— arguably Section 6.2 is the chief component of this present work.

If for a sample s at hand of Selection Units when determining A_j for j in s and the mutually interlinked m_i values, the required empirical work grows up prohibitively, let us proceed as follows:

Let C_j be the number of Observation Units in A_j. Then $C = \sum_{j \in s} C_j = C(s)$, say, is the total number of Observation Units to be surveyed.

In case C seems too large jeopardizing an effective implementation of a sample survey as planned, then we recommend independently choosing by Simple Random Sampling Without Replacement method sub-samples B_j from each respective A_j for j in s of sizes d_j out of C_j for $2 \leq d_j \leq C_j$ such that $D = \sum_{j \in s} d_j = D(s)$, say, is of a manageable magnitude. Then

$B(s) = \bigcup_{j \in s} B_j$ is the 'constrained sample size network sample' derived from $A(s) = \bigcup_{j \in s} A_j$, the initial network sample.

Then a revised unbiased estimator for W is

$$e = \sum_m \frac{Q_u}{p_u} \left(\frac{C_u}{d_u} \sum_{i \in B_u} \frac{y_i}{m_i} \right) = \sum_m \frac{Q_u}{p_u} \psi_u, \text{ say,}$$

writing

$$\psi_u = \left(\frac{C_u}{d_u} \sum_{i \in B_u} \frac{y_i}{m_i} \right).$$

It follows that given the initial $A(s)$, the conditional expectation of ψ_u is w_u, because we may write $E_C(\psi_u | A(s)) = w_u$ and so $E_C(e | A(s)) = t = t(s)$. So, e is an unbiased estimator for W.

It follows that

$$v_c(\psi_u) = \left(\frac{1}{d_u} - \frac{1}{C_u}\right)\left(\frac{1}{d_u - 1}\right)\sum_{i \in B_u}\left(\frac{y_i}{m_i} - \frac{1}{d_u}\sum_{i \in B_u}\frac{y_i}{m_i}\right)^2$$

is an unbiased estimator for the conditional variance of ψ_u given $A(s)$, namely $V_C(\psi_u)$.

We may observe $E = E_p E_C$ and $V = E_p V_C + V_p E_C$, denoting by E_C, V_C the conditional expectation and variance operators given $A(s)$, E_p, V_p the expectation, variance operators in respect of Rao, Hartley & Cochran's sampling scheme and by E, V the overall expectation and variance operators.

Thus

$$E(e) = E_p\left[E_C\left(\sum_m \frac{Q_u}{P_u}\psi_u\right)\right] = E_p\sum_n \frac{Q_u}{P_u}w_u$$

$$= E_p(t) = W = Y,$$

$$V(e) = V_p(t) + E_p\left[\sum_m \left(\frac{Q_u}{P_u}\right)^2 V_C(\psi_u)\right],$$

Noting $V_C(\psi_u) = C_u^2\left(\frac{1}{d_u} - \frac{1}{C_u}\right)\left[\sum_{i \in A_u}\left(a_i - \frac{1}{C_u}\sum_{i \in A_u}a_i\right)^2\right]\left(\frac{1}{C_u - 1}\right)$

writing $a_i = \frac{y_i}{m_i}$, $i \in A_u$ and recalling

$$V_p(t) = A\left[\sum_{j=1}^M \frac{w_j^2}{P_j} - W^2\right] = A\sum_{j<}^M\sum_{j=1}^M P_j P_{j'}\left(\frac{w_j}{P_j} - \frac{w_{j'}}{P_{j'}}\right)^2$$

and verifying that

$$v_p(t) = B\sum_m\sum_m Q_u Q_{u'}\left(\frac{w_u}{P_u} - \frac{w_{u'}}{P_{u'}}\right)^2$$

satisfies $E_p\, v_p\,(t) = V_p(t)$ and $E_p\, v_C\,(\psi_u) = V_C\,(\psi_u)$, let us derive as follows an unbiased estimator $v(e)$ for $V(e)$.

Let
$$G = B\sum_m\sum_m Q_u Q_{u'}\left(\frac{\psi_u}{p_u} - \frac{\psi_{u'}}{p_{u'}}\right)^2$$

and
$$H = \sum_m\left(\frac{Q_u}{p_u}\right)^2 v_c(\psi_u).$$

Then,
$$E_C\, G = B\sum_m\sum_m Q_u Q_{u'}\left[\left(\frac{w_u}{p_u} - \frac{w_{u'}}{p_{u'}}\right) + \frac{V_C(\psi_u)}{p_u^2} + \frac{V_C(\psi_{u'})}{p_{u'}^2}\right]$$

$$= v_p\, t\; + B\sum_m\sum_m Q_u Q_{u'}\left[E_C\left(\frac{V_C\,\psi_u)}{p_u^2} + \frac{V_C(\psi_{u'})}{p_{u'}^2}\right)\right]$$

and
$$E_C(H) = \sum_m \frac{Q_u^2}{P_u^2} V_C(\psi_u).$$

So,
$$v(e) = B\sum_m\sum_m Q_u Q_{u'}\left[\left(\frac{\psi_u}{p_u} - \frac{\psi_{u'}}{p_{u'}}\right)^2 - \left(\frac{V_C(\psi_u)}{p_u^2} + \frac{V_C(\psi_{u'})}{p_{u'}^2}\right)\right]$$

$$+ \sum_m\left(\frac{Q_u}{P_u}\right)^2 v_c\left(\psi_u\right)$$

has $E[v_e] = V(e)$

A comparable formula for $v(e)$ as
$$(1+B)\sum_m v_c(\psi_u)\left(\frac{Q_u}{p_u}\right)^2 + B\left(\sum \psi_u^2 \frac{Q_u}{p_u} - e^2\right)$$

was earlier derived in a slightly different context by Chaudhuri (2003).

Turning now to $t' = \dfrac{M}{m} \sum_{j \in s} w_j$, similarly keeping everything else intact but replacing RHC scheme above by Simple Random Sampling Without Replacement of m Selection Units out of M Selection Units, we have the following counter-parts derived, corresponding to t above retaining the notations:

$$V_p(t') = M^2 \left(\frac{1}{m} - \frac{1}{M} \right) \frac{\displaystyle\sum_{j=1}^{M} \left(w_j - \frac{\displaystyle\sum_{1}^{M} w_j}{M} \right)^2}{(M-1)},$$

$$v_p(t') = M^2 \left(\frac{1}{m} - \frac{1}{M} \right) \frac{\displaystyle\sum_{j \in s} \left(w_j - \frac{\displaystyle\sum_{j \in s} w_j}{m} \right)^2}{(m-1)},$$

$$\psi_j = \frac{C_j}{d_j} \sum_{i \in B_j} a_i, \; a_i = \frac{y_i}{m_i}, \; B_j \subset A_j, \; e' = \frac{M}{m} \sum_{j \in s} \psi_j.$$

Then

$$E_C(\psi_j) = w_j.$$

$$E_C(e') = t'.$$

$$V(e') = V_p(t') + E_p(V_C(e')).$$

Now,

$$V_C(\psi_j) = C_j^2 \left(\frac{1}{d_j} - \frac{1}{C_j} \right) \sum_{i \in A_j} \left(a_i - \frac{\displaystyle\sum_{i \in A_j} a_i}{C_j} \right)^2 \Bigg/ (C_j - 1)$$

$$V_C(e') = \left(\frac{M}{m} \right)^2 \sum_{j \in s} V_C(\psi_j),$$

$$v_C(\psi_j) = C_j^2 \left(\frac{1}{d_j} - \frac{1}{C_j} \right) \sum_{i \in B_j} \left(a_i - \frac{\sum_{i \in B_j} a_i}{d_j} \right)^2 \Bigg/ (d_j - 1),$$

$$v_C(e') = \left(\frac{M}{m} \right)^2 \sum_{j \in s} v_C(\psi_j)$$

so that

$$E_C v_C(e') = V_C(e').$$

So,

$$V(e') = V_p \left(\frac{M}{m} \sum_{j \in s} w_j \right) + E_p E_C [v_C(e')]$$

$$= M^2 \left(\frac{1}{m} - \frac{1}{M} \right) \frac{\sum_{j=1}^{M} \left(w_j - \frac{\sum_{j=1}^{M} w_j}{M} \right)^2}{(M-1)}$$

$$+ E_p E_C \left[\left(\frac{M}{m} \right)^2 \sum_{j \in s} v_C(\psi_j) \right]$$

$$= E_p M^2 \left(\frac{1}{m} - \frac{1}{M} \right) \frac{\sum_{j \in s} \left(w_j - \frac{\sum_{j \in s} w_j}{m} \right)^2}{(m-1)}$$

$$+ E_p E_C \left(\frac{M}{m} \right)^2 \sum_{j \in s} C_j^2 \left(\frac{1}{d_j} - \frac{1}{C_j} \right) \frac{\sum_{i \in B_j} \left(a_i - \frac{\sum_{i \in B_j} a_i}{d_j} \right)^2}{(d_j - 1)}.$$

Letting

$$
G' = \frac{\sum\limits_{i \in s} \left(\psi_j - \dfrac{\sum\limits_{i \in s} \psi_j}{m} \right)^2}{m - 1}
$$

we get

$$
E_C(G') = \frac{1}{(m-1)} \left[\sum_{j \in s} \left(w_j - \frac{\sum\limits_{j \in s} w_j}{m} \right)^2 + \sum_{j \in s} v_C(\psi_j) \right]
$$

Consequently,

$$
v(e') = \frac{M^2 \left(\dfrac{1}{m} - \dfrac{1}{M} \right)}{(m-1)} \left[\sum_{j \in s} \left\{ \left(\psi_j - \frac{\sum\limits_{j \in s} \psi_j}{m} \right)^2 - v_C(\psi_j) \right\} \right]
$$

$$
+ \left(\frac{M}{m} \right)^2 \sum_{j \in s} C_j^2 \left(\frac{1}{d_j} - \frac{1}{C_j} \right) \frac{\sum\limits_{i \in B_j} \left(a_i - \dfrac{\sum\limits_{i \in B_j} a_i}{d_j} \right)^2}{(d_j - 1)}
$$

satisfies $Ev(e') = E_p \, E_C \, [v(e')] = V(e')$ and hence provides an unbiased estimator for $V(e')$.

Thus $v(e')$ simplifies clearly to

$$
v(e') = \frac{M^2 \left(\dfrac{1}{m} - \dfrac{1}{M} \right)}{(m-1)} \sum_{j \in s} \left(\psi_j - \frac{\sum\limits_{j \in s} \psi_j}{m} \right)^2
$$

$$
- \frac{M(M-m)}{m(m-1)} \sum_{j \in s} C_j^2 \left(\frac{1}{d_j} - \frac{1}{C_j} \right) \frac{\sum\limits_{i \in B_j} \left(a_i - \dfrac{\sum\limits_{i \in B_j} a_i}{d_j} \right)^2}{(d_j - 1)}.
$$

6.2.4 Estimating Total Household Medical Expenses by Sampling Hospital In-patients

In a study conducted in the West Tripura district of North-East India, we took hospitals, nursing homes, clinics, primary health centers, community health centres as the Selection Units; and the Households with patients with heart diseases, cancer or gall bladder treated in any of them as the Observation Units, both within the West Tripura district, India, we adopted "constrained sample-size network sampling" to obtain unbiased estimate of the total health expenses of the households of this district. Obviously Households with no patients were excluded. Our reference period was a year for which any recorded in-patient's household was traced to establish the "Selection Unit-Observation Units links". The total number of beds in a "Treatment House" was taken as a size measure of a Selection Unit. Other details are omitted. Once the total of a variable of interest was estimated without a bias and (1) the total number of such Observation Units using an indicator function we (2) appropriately use a ratio estimator, vide Chaudhuri (2010) for their per capita household expenses in the district. Even household expenses over the treatment of the inmates by age-groups are also easily estimated using right indicator functions. Since Coefficients of Variations are the performance characteristics the tables below briefly narrate our real-life findings. The relevant estimation procedural details are briefly narrated in the Appendix which the reader is requested to refer to.

TABLE 6.1 Age, Sex and Disease-wise observed distribution of the households sampled

Age (years) of in-patient treated	Heart Disease (H)		
	Male	Female	Total
≤40	15	15	30
41-50	25	12	37
51-60	35	17	52
61-	31	34	65
Total	106	78	184
	Cancer(C)		
	Male	Female	Total
≤40	14	2	16
41-50	28	18	46
51-60	55	19	74
61-	98	15	113
Total	195	54	249

(Contd.)

TABLE 6.1   Age, Sex and Disease-wise observed distribution of the households sampled (*Contd.*)

Age (years) of in-patient treated	Heart Disease (H)		
	Male	Female	Total
	Gall Bladder (G)		
	Male	Female	Total
≤40	7	63	70
41-50	16	66	82
51-60	35	53	88
61-	8	23	31
Total	66	205	271

TABLE 6.2   Estimated Percentage Distribution of the in-patients treated in Healthcare Centres in West Tripura District age, sex and disease-wise for the actual sample-size of the Households (HH) visited and surveyed

	Heart Disease (H)								
	Male			Female			Total		
Age (years) of in-patient treated	Esti-mate (%)	cv (%)	Sample size of HH's visited	Esti-mate (%)	cv (%)	Sample size of HH's visited	Esti-mate (%)	cv (%)	Sample size of HH's visited
≤40	23.1	34.4	15	20.8	14.9	15	22.7	21.0	30
41-50	22.0	28.0	25	20.0	22.9	12	21.0	23.6	37
51-60	32.2	18.1	35	15.7	30.1	17	26.7	15.6	52
61-	22.7	22.4	31	43.5	9.1	34	29.6	13.5	65
Total	100.0		106	100.0		78	100.0		184
	Cancer (C)								
	Male			Female			Total		
≤40	6.4	22.6	14	3.7	54.4	2	5.6	26.9	16
41-50	14.8	14.4	28	35.2	15.1	18	19.3	10.8	46
51-60	29.5	9.1	55	37.0	14.5	19	31.1	7.9	74
61-	49.3	5.8	98	24.1	19.2	15	44.0	5.7	113
Total	100.0		195	100.0		54	100.0		249
	Gall Bladder Disease (G)								
	Male			Female			Total		
≤40	10.7	35.7	7	33.4	13.0	63	30.2	15.2	70
41-50	14.8	44.6	16	27.8	19.5	66	25.1	14.2	82
51-60	60.8	11.7	35	19.5	20.2	53	26.2	15.6	88
61-	13.7	28.6	8	19.3	28.3	23	18.5	25.3	31
Total	100.0		66	100.0		205	100.0		271

Table 6.3 **Estimated per Household expenses during 1 Nov. 2004 – 31 Oct. 2005 for treatment in Healthcare centres**

Heart Disease(H)				
	Rural	**Urban**	**Total**	**Total sample size**
Estimated expenses (₹)	33668	57084	44267	
Coefficients of Variation (%)	23.7	22.1	18.5	
	Hindu	**Muslim**	**Others**	
Estimated expenses (₹)	44652	12496	46320	
	Schedule Caste	**Schedule Tribe**	**Others**	
Estimated expenses (₹)	34867	25463	45875	184
Heart Disease(H)				
	Rural	**Urban**	**Total**	**Total sample size**
Estimated expenses (₹)	75440	89214	79912	
Coefficients of Variation (%)	7.9	12.1	5.8	
	Hindu	**Muslim**	**Others**	
Estimated expenses (₹)	84032	37704	48915	
	Schedule Caste	**Schedule Tribe**	**Others**	
Estimated expenses (₹)	64602	30697	84595	244
Heart Disease(H)				
	Rural	**Urban**	**Total**	**Total sample size**
Estimated expenses (₹)	12215	13193	12585	
Coefficients of Variation (%)	21.5	19.9	19.9	
	Hindu	**Muslim**	**Others**	
Estimated expenses (₹)	12627	3935		
	Schedule Caste	**Schedule Tribe**	**Others**	
Estimated expenses (₹)	14051	5074	12348	270

6.2.5 Concluding Remarks

We set a target of covering a total of 700 households in the West Tripura District. But we could actually get response from 698 households in all. This is clear from Table 6.2. But Table 6.1

accounts for a total of 704 households claimed to be covered. This is not an enigma. We gathered data on 704 household members who received treatments as in-patients or heart disease, cancer and/or gall bladder irregularities altogether out of a total of 698 households. No contradiction is apparent.

Among those treated for cancer only about 22% are females and 41% are females who were also treated for heart related diseases. But this picture is reversed in case of gall bladder — here 76% are females. In the younger ages, females treated for gall bladder far outnumber the male household members. These relate to the facts experienced for the sampled households only.

When we turn to estimation for the entire district, we find that among those aged 50-60 years, half of the people are women while among those who are elderly still, half of the people are men — this is among those who suffer from heart disease. Among the cancer patients treated in healthcare institutions the female/male ratios are about 1/2 among those aged 50-60 years and 2/5 among those aged higher still respectively. Among the gall bladder patients these ratios are 1/3 and 1:1 roughly in these two age-groups.

Per household expenses for treatment of cancer is quite huge both in urban and rural areas, not quite worthy of attention the same for heart disease and for gall bladder treatment. The expenses are not great enough to cause any serious alarm. The expenses, however, vary significantly across villages vs. cities, Hindus vs. Muslims and caste-wise. About the accuracies in estimation we may only mention that except when the sample-size is too small the coefficient of variation is quite low in magnitude indicating desirable accuracy level.

Possibly, contrary to popular anticipation even in case of treatment of cancer the costs incurred within the particular district exceed those outside. But incidental expenses incurred outside far exceed those within, possibly for travel costs of the patients and their attendants. For heart treatment also, the expenses incurred outside the district are quite less compared with those inside. But the same is the case for those on incidental items. This probably implies that for treatment of heart disease people do not greatly move out.

For treatment of gall bladder problems probably the people do not seek attention from outside the district. Also incidental expenses vis-à-vis those for actual treatment are rather insignificant.

6.3 AN EXERCISE IN MATERNAL MORTALITY ASSESSMENT IN A DISTRICT

Maternal mortality is a sensitive indicator of health and environment as well as socio-economic conditions in which people live. A maternal death is one of a woman in pregnancy or within 42 days of termination of pregnancy from any cause related to its aggravation or mismanagement but not due to any other unrelated causes or accidents. About 20% of the world's total population is exposed to the risk of repeated pregnancy, the main causes of maternal deaths are tetanus, anaemia, toxemia, obstructive labour complications in abortion, infection and haemorrage, pregnancy at too early or advanced ages, wrong spacing and other shortcomings. World's conscience is often shaken by phenomena of too high frequencies of maternal deaths especially in under-developed as well as developing countries.

In *Demography India*. Vol. 31, pp. 253–58 published in 2002, the author and his colleagues headed by Professors M. Pal and P. Bharati have noted that in their concerned study at the end of the last century, there used to take place 1 in 48 women in the child-bearing ages 15-49 last birthday (lbd) running the risk of puerperal death in developing countries while in developed countries the incidence rate was a mere '1' in 1800 such women.

Since it is a common knowledge that maternal deaths are more abundant in the rural areas compared to the urban areas, Indian Statistical Institute (ISI) undertook a UNICEF-sponsored project vide Bose (1998) entitled "Estimation of Infant and Maternal Mortality Rates Identification of Their Determinants in North 24 Parganas, West Bengal" to study among other things, incidences of maternal deaths and their circumstances in an extensive rural area in late 1998. The author happened to put in his efforts to the collective ones by a team headed by Professors P. Bharati, M. Pal and B.N. Bhattacharya. They considered the twin concepts of Maternal Mortality namely (i) Maternal Mortality Rate (MMR) which is the "number of maternal deaths per 100,000 mothers" and (ii) Maternal Mortality Ratio which is the "number of maternal deaths per 100,000 live births".

In obtaining both (i) & (ii), (a) the Adaptive Sampling method was used covering informants including relatives other than sisters of initially encountered mothers surveyed in their respective addresses determined from the former and (b) in addition the "sisterhood" method which includes only the relevant sisters but not other women in the child-bearing ages as referred to by the initially sampled mothers encountered.

Since no numerical figures are presented here, the reader's attention is solicited to the Appendix presented at the end of this monograph showing how 'Network Sampling' with constrained sample size may be employed in practice.

6.4 ADAPTIVE AND NETWORK SAMPLING FOR A DRUG ABUSE SURVEY

Since time immemorial, mankind has shown a very widely observable inclination for consuming 'mood altering' substances. Initially they were herbal and natural products on which refinements, through human efforts started being exercised. But gradually over time sophistications started being exerted with rapid enhancement in intensity. So most of the currently used items are synthetic products with various levels of technological advances in inducements on their predecessors.

A few popular ones are:

- Heroin (Brown sugar, Smack, No. 4, Junk, etc.)
- Opium (Post, Reta, Poppy, etc.)
- Buprenorphine (Norphine, TDX, Tidigesic, Adnok, etc.)
- Other opiates like Pethidine, Pentazocine, Fortwin, Codeine, Methaone, etc.
- Cannabis (Ganja, Charas, Hashish, Grass, Sulfa, etc.)
- Alcohol
- Barbiturates, like Phenobarbitore, Pentobarbitore, Gardanal, Seconal, etc.
- Minor tranquilizers such as Valium, Diazepam, Librium, Nitravart, Das Number ki Goli, Alprax, etc.
- Other sedatives/hypnotics like Sedyn, Carisoma, Mandrax, etc.;
- Cocaine like Crack
- Amphetamine also described as Speed, Dextroamphet-amine, Ecstasyate;
- Hallucinogens like PCP, Angel, Dust, LSD, etc.
- Inhalants like Glue, Petrol, Kerosene, Thinner Solvate, Paints, Araldite, Quickfix, Spirit, etc.;

- Others like Avil, Vrufen, Voveran, Laxative, Lodex, Boot polish, etc.

To these can be added the substances injected into the human system using syringes. Unregulated consumption of these substances induces intoxicating obsessions among people often leading to malevolent social, personal, sexual and family specific behavior patterns that one may recognize as socially derogatory so as to characterize these as " Drug Abuses".

In developing and underdeveloped countries, these abuses are more rampant. United Nation's International Drug Control Programme takes formidable steps from time to time, taking stock of the prevailing circumstances across various such countries.

In India, the Ministry of Social Justice and Empowerment takes cognizance of this as an issue of national importance. But till around 2000 data gathered on drug abuse were scanty, through epidemiological studies held in a few scattered sites only.

"National Household Survey of Drug and Alcohol Abuse in India" undertaken by Professor Anurag Srivastava of the Clinical Epidemiology Unit of the All India Institute of Medical Sciences, New Delhi was the first empirical investigation executed in India at the national level (in 2000-2001). In this study Indian Council of Medical Research, Ministry of Social Justice and Empowerment and National Justice for Social Defence were also involved. This received guidance from the *World Drug Report*, 1997 as a major source of background. The following data from here was duly taken note of by these investigators:

Table 6.4 Estimated Number of Drug Abuses Worldwide (Annual Prevalence)

Substance	Percentage of total population
Alcohol	50%
Tobacco	20%
Any illicit drug	10%
Cannabis	2.5%
Sedatives	4%
Stimulants	0.5%
Hallucinogens	0.44%
Opiates (including Heroin)	0.14%
Cocaine	0.23%
Any drug through injection route	0.05%

Source: World Drug Report, (1997).

World Drug Report 2000 notes some changes hereupon, The 1997 NHSDA (National Household Survey on Drug Abuses) of USA observed that out of 216 million people studied, most commonly used illicit drug was marijuana (cannabis) with cocaine and heroin ranking second and third respectively. Among licit substances, 82% of the people ever used alcohol while 71% people consumed tobacco, smoked and unsmoked.

In 2007 the Ministry of Statistics and Programme Implementation of the Government of India under a Ministry of State constituted a Committee to study the feasibility of conducting a pilot survey of drug abuse with the help of certain Non-government Organizations under the guidance of the National Sample Survey Organization (now called Office NSSO). At the outset, this committee observed that "The magnitude of drug abuse and its correlations in the realm of socio-demographic traits has not been well-researched in India. Probably the country is too large and the resources too limited. Hence the deficiency in the data-contents. Following the pioneering exercise by Ministry of Social Justice & Empowerment, in collaboration with the United Nations International Drug Control Programme in 2007, Ministry of Social Justice and Empowerment sought the assistance from National Sample Survey Office this time to conduct a pilot survey in limited areas on drug abuse after receiving collaboration from the Central Statistical Organization (now called "Office"), of the Government of India and certain Non-Government Organizations active in these areas locally in several territorial segments of India. Assistance was offered to National Sample Survey Office and Central Statistical Office by the All India Institute of Medical Sciences, United Nations Office on Drug and Crime, and the Regional Office for South Asia among others. A pre-testing of the draft questionnaires was held in three selected villages and (i) listing schedule and main schedules of enquiry were canvassed in households. This debut was not a happy one of a 'Drug Survey' though as a pilot, it had consensus of the pioneering organizers. At the instance of Ministry of Social Justice & Empowerment, the National Sample Survey Office agreed to do the needful to generate national and state-wise estimates of the extent of alcohol abuse and other substances for the target population of all people in India aged 12-65 (males and females). For this a Steering Committee for National Sample Survey in its meeting of 13-14 February 2008 decided to form a Committee with the present author as its Chairman to suggest appropriate survey methodology and schedules of enquiry for undertaking pilot surveys on drug abuse. A requirement was spelled out for devising estimation methods for the number of users of hard drugs (e.g. heroin, cocaine, etc.) or injecting drugs so as to help policy-makers, as regards:

(i)   Planning and allocation of resources for control, treatment and prevention of problem drug use and its consequences,

(ii) Monitoring key targets of drug policy,

(iii) Interpreting impact and socio-economic implications associated with use of drugs and

(iv) Profiling the database in varying dimensions such as age, sex, region, socio-economic hierarchy, etc. for addressing the issue in a focused manner.

The core problem identified was to find a scientific method for estimating the number of users of heroin, crack, cocaine and other harmful substances.

Hard drug users are, however, hard to access. Standard survey procedures fail to identity them mainly because (1) they escape customary frames like household groups, (2) habitual under-reporting owing to social stigma, and (3) relative rarity of such cases. So, alternative strategies are to be tried like (a) Capture-Release-Recapture, (b) Snowball Sampling, (c) Adaptive Sampling (d) Network Sampling and (e) Randomized Response Techniques as opined by the Committee comprising this author and others set up in February-March 2008.

In the technique (a) above, first a Simple Random Sample Without Replacement is chosen, each unit is given, marked with an indelible ink and the entire sample is returned to the population after a reasonable lapse of time. Then a second Simple Random Sample Without Replacement is taken. Writing n_1, n_2 as the sizes of these two Simple Random Samples Without Replacement and K as the number of units marked in the first but recaptured in the second, the integer part of the number $\frac{n_1 n_2}{K}$ provides the Maximum Likelihood Estimate for the population size. This may provide a basis for an intelligent guess about the unknown total number of drug users of various categories in different communities. It was Feller (1977) who initiated this theory.

The Snowball Sampling introduced by Goodman (1961) starts with an arbitrary probability sample. Every unit is asked to give the whereabouts of his/her best K friends ($K \geq 1$) who constitute a sample in the stage 1, each of whom provides the location in turn of his/her best K (≥ 1) friends giving the sample in stage 2 and so on. Some committee members find this scheme interesting. The present author is not impressed as this "Snowball" sample in large-scale surveys does not seem feasible in yielding serviceable estimates affording measures of errors. Adaptive and Network sampling are the principal themes in this treatise to be discussed further in depth.

Randomized Response Technique was introduced by Warner (1965) with latest developments expounded by Chaudhuri (2011) on devices to generate feasible data relating to sensitive and stigmatizing issues. For example, adopting only sampling scheme admitting positive inclusion—probabilities of every single and paired population members taking hold of a selected individual may do the following:

Suppose a member bears a sensitive trait A or its complement A^C and an innocuous characteristic B or its complement B^C, the former being called a variable y and the latter an unrelated variable x, say, such that for a person labelled i,

$$y_i = 1 \text{ if } i \text{ bears } A$$
$$= 0 \text{ if } i \text{ bears } A^C,$$
$$x_i = 1 \text{ if } i \text{ bears } B$$
$$= 0 \text{ if } i \text{ bears } B^C.$$

Then to estimate the population total of y_i's an investigator may approach a sampled person i with 2 boxes containing cards marked A, B in proportions

$$p_1 : (1 - p_1),\ 0 < p_1 < 1,\ \sum p_1 = 1$$

and the second box containing A, B in proportions

$$p_2 : (1 - p_2),\ 0 < p_2 \neq p_1 < 1.$$

The person addressed is to randomly choose one card from the 1st box responding

$$I_i = 1 \text{ if card type matches the trait}$$
$$= 0, \text{ else}$$

and $J_i = 1/0$, similarly for a draw from the second box.

Then,

$$r_i = \frac{(1-p_2)I_i - (1-p_1)J_i}{(p_1 - p_2)}$$

unbiasedly estimates y_i and $r_i(r_i - 1)$ unbiasedly estimates V_i, the variance of r_i in respect of this Randomized Response generation procedure.

Now following the usual principle of two-stage sampling procedure, the total of y_i's is easily unbiasedly estimated with an unbiased variance estimator easily derived. A theory is developed elegantly as well covering arbitrarily real valued variables in estimating their totals and means.

In 2008–09, a pre-testing with alternative survey methodologies was tried in four specified sites in Kolkata. Network-cum-Adaptive sampling design was adopted. Only 35 drug addicts in two days could be interviewed by National Sample Survey Organization in collaboration with two Non-Government Organizations giving us interesting profiles for themselves. Adapting sampling could not be executed with success. But Network sampling was impressive in identifying hard-core addicts.

A pilot study on drug addicts was done in three municipalities in three cities of three different Indian states. The following stratified sampling scheme for such surveys in each of the Indian states and Union Territories was planned.

The stratum 1 of all drug-related treatment/correction centres and de-addiction centres within a state/Union Territories was planned to be surveyed by constrained network sampling separately in two sub-strata 11 and 12.

The sub-stratum 11 excluded and sub-stratum 12 included the big million-plus populated cities in a state/Union Territory.

Stratum 2 composed of sub-stratum 21 and 22 respectively excluding and including the big million-plus cities consist of all pockets of consumption and trade centres of drugs in the districts sampled from the states/Union Territories. Stratified constrained network sampling is recommended for stratum 22. For the sub-stratum 21 the Rao, Hartley and Cochran's scheme in the first stage and constrained network sampling in the second stage are planned. Stratum 3 consists of all the red-light areas in the districts suitably probability-sampled from the states/Union Territoriess. The sub-stratum 31 excludes the million-plus populated cities but sub-stratum 32 includes only the smaller cities. In sub-stratum 31 Rao, Hartley and Cochran's sampling scheme is used in the first stage and constrained network sampling in the second stage, and similarly in sub-stratum 32. Stratum 4 consists of all the prisons/correction centres within a state/UT. State level and relational estimates are then to be developed by amalgamation. It is gratifying to observe that the authorities concerned in this project could be brought to the general view that a drug use survey is inherently distinct from usual social surveys demanding anonymity and protected privacy in divulging secrets and behavioral idiosyncrasies.

In early 2007, the Ministry of Social Justice and Empowerment, Government of India requested National Sample Survey Office to generate by dint of an appropriate sample survey, estimates concerning the target Indian population aged 12-65 years, both male and female, concerning the following items:

1. Extent of alcohol abuse and other substances in India,
2. Pattern and trends of alcohol and drug abuse ever since 2000
3. Drug consumption among adolescents and youth
4. Prevalence of infections like HIV/AIDS, Hepatitis B & C, STD's among them
5. Injecting Drug Users and other drug users
6. Prevalence of drug use among school children
7. Modes of drug intake, whether oral, inhalation, Injecting Drug Users, etc.
8. Sexual practices, diseases and treatment history of drug users
9. Legal situation
10. Socio-economic and health situation of drug users, etc.

Unfortunately, though National Sample Survey Organization forwarded to Ministry of Social Justice and Empowerment, the sponsor, the results of the pre-lists and the Pilot survey along with its proposal for the methods to be tried in various states/Union Territories in 2010, green signal has not been received so far.

7

A Brief Review of Available Literature

Abstract

Thompson (1990,1992) and Thompson & Seber (1996) are the pioneering source materials inducing our interest and zeal as reflected in our debut write-up in Chaudhuri (2000) on the twin subjects (i) Network Sampling and (ii) Adaptive Sampling. The subject Adaptive Sampling is developing fast as summarized, reviewed and intensified by Seber & Salehi (2013). For reasons not known the coverage started and is progressing clumsily with involved algebra and concepts as well. To our belief, the literature that grew as following Chaudhuri's (2000) in the hands of the present author's colleagues and collaborators namely Chaudhuri, Bose & Ghosh (2004), Chaudhuri & Pal (2005), Chaudhuri & Saha (2004), Chaudhuri, Bose & Dihidar (2004) and Chaudhuri, Bose & Ghosh (2005) and also as ventured in the current monograph, appears appreciably rather simple and reader-friendly. It is briefly recounted in this chapter.

7.1 CERTAIN DETAILS

7.1.1 Chaudhuri, Bose & Ghosh's (2004) Applications

A typically Indian problem of developing reliable estimates of Gross Domestic Products because of rural preponderance of small-scale industries in the unorganized sector, heightened by difficulties of locating the workers engaged therein and hence ascertaining their numbers, was addressed by Chaudhuri, Bose & Ghosh (2004). They concentrated on a particular neighbouring district of Kolkata and utilized the available survey results concerning it to initiate a stratified two-stage equal as well as unequal probability sample selection so as to estimate the district totals of people earning their livelihoods through occupations in 10 rural rudimentary industries such as

   (i) Handloom (H),
  (ii) Bamboo (B),
 (iii) Husking (HU),
  (iv) Poultry (P),
   (v) Silk (S),
  (vi) Stone-breaking (SB),
 (vii) Bidi-making (BM),
(viii) Ironsmithy,
  (ix) Carpentry (C) and
   (x) Paddy-crushing (PC).

The message brought out by their theoretical-cum-empirical exercise is that (a) traditional sampling procedure takes a backseat behind (b) Adaptive Sampling and (c) defining neighbourhood in network formation is of crucial importance.

7.1.2 Chaudhuri & Pal's (2005) Approach

Chaudhuri & Pal (2005) consider estimating the total unknown areas under "Mining and Industrial Wastelands" in various districts in four different regional territories of India. They first take Rao-Hartley-Cochran samples and employ 'Small Area Estimation' procedures following the principle of Model Assisted Sampling, then empirical Bayesian modification and also extending the initial sample to an

Adaptive Sample demonstrating the advantages gained in efficient estimation.

For a quick understanding let us briefly clarify the concepts involved. A population U of N units is often divisible into mutually disjointed parts U_d, $d = 1,\ldots, D$ called domains and estimation of the total $Y = \sum_1^N y_i$ of a variate y with unit-values y_i, $i \in U$ as well as of the domain totals $Y_d = \sum_{i \in U_d} y_i$, $d = 1, 2,\ldots, D$ is occasionally demanded. Even if the total size n of a sample s taken from U is large, the sample-size n_d hailing from U_d may be inadequate in achieving satisfactory results for the estimation of the domain totals. Then arises the problem of "Developing Small Domains Statistics", called Small Area Estimation in case U is territorial/geographical. A possible way out is "borrowing" strength using additional sample-observations outside the domain but inside several "like domains" in the sample with suitably postulated models connecting the various domains through regression relationships with domain-wise common parameters of relevance.When forms of estimators are suggested by tried and tested models but their performance characteristics are taken honouring the "Design-based" traditional principles, we say we are adopting the "Model Assisted" version. Accordingly, Chaudhuri & Pal (2005) consider for Y_d the 'Synthetic Generalized Regression' estimator

$$t_{sgd} = t_d(Y) + b_Q(X_d - t_d(x)).$$

Here,

$$t_d(y), t_d(x)$$

are traditional design-based estimators for Y_d and

$$X_d = \sum_{i \in U_d} x_i,$$

x_i, the value of an available x related to y for i in U and

$$b_Q = \frac{\sum_{i \in s} y_i x_i Q_i}{\sum_{i \in s} x_i^2 Q_i}$$

with Q_i as suitably chosen positive numbers.

A regression model

$$y_i = \beta x_i + \epsilon_i, i \in U$$

is postulated with a common slope β for the entire population and errors ϵ_i simplistically independent in nature.

Further postulating $\hat{Y}_d = t_{sgd}$ as distributed conditionally, given Y_d, as normal with mean Y_d and variance same as the estimated variance of $\hat{Y}_d$ and Y_d as also normally distributed, one may take the conditional expectation of Y_d. Recognizing unknowable parameters in this, an empirical Bayes estimator then is derivable substituting suitable estimators for these parameters. Prasad & Rao (1990) and Rao & Lahiri (1995) provide Mean Square Error estimators for them. Chaudhuri & Pal (2005) examine simulation-based efficacy levels of the synthetic greg and empirical Bayes estimators. Finally they show how adaptive samples strengthen their claims for acceptability. Bayes and empirical Bayes estimators are purely model-dependent. But a model assisted estimator-predictor is only motivated by a model and is robust because of its design-based properties which hold despite model failures.

7.1.3   Chaudhuri & Saha's (2004) Findings

In India Economic Censuses are intermittently held at approximately 5 yearly intervals in order, chiefly, to ascertain nation-wide numbers regionally of earners through diverse non-agricultural pursuits. These exercises not only produce contemporary figures of immediate uses but also help planning future surveys in effective ways. Utilizing Indian Economic Censuses figures gathered in 1998, Chaudhuri & Pal (2005) illustrate empirically through simulations comparative performances of (I) Two-stage Rao-Hartley-Cochran's sample selection employing Generalized Regression estimators, (II) Modification of Rao-Hartley-Cochran's initial sampling to its Adaptive extensions and comparison of both versus (III) traditional single-stage cluster sampling at random without replacement. They address the problem of estimating inclustry-wise rural workforce exploring alternative ways of defining Network, noticing their relative efficiencies. Importance of Adaptive Sampling is vindicated by their numerical exercise as demonstrated in one location cited.

Notably, Chaudhuri & Pal (2005) recognize the "Asymptotic Design Consistency" and equivalently "Asymptotic Design Unbiasedness" of Generalized Regression estimators (predictors in an equivalent sense) adopting Brewer's (1979) approach. In order to modify Cochran's (1953, 1963, 1977) concept of Finite Consistency applicable to finite survey population situations, according to which a statistic t is finitely consistent for a parameter if it coincides with the latter in case the sample coincides with a population or more loosely if a sample-size matches the population size. Brewer(1979) introduced his new limiting concepts in sample surveys. He supposed that one may hypothesize an infinite sequence of finite survey populations U_K, $K = 1,\ldots, \infty$ from each of which samples s_K, $K = 1,\ldots, \infty$ may conceptually be drawn independently according to suitable designs. Then a statistic t_K based on s_K may be examined on applying Chebyshev's inequality to converge to its expected value, as K tends to infinity, thus being Asymptotic Design Consistent and simultaneously Asymptotic Design Unbiased as well. In application, Slutzky's theorem vide Cramer (1946) is helpful in testing for this. For example, the greg predictor of the form

$$t_g = t(y) + b_Q (X - t(x))$$

with t_y, t_x as unbiased estimators for

$$Y = \sum_1^N y_i , \quad X = \sum_1^N x_i$$

and

$$b_Q = \frac{\sum_s y_i x_i Q_i}{\sum_s x_i^2 Q_i}, \quad Q_i > 0 \,\forall\, i,$$

is Asymptotic Design Consistent and Asymptotic Design Unbiased for Y. This is so because b_Q and t_g are well-behaved rational functions so that for large samples, while Slutzky's theorem says that if u_n, v_n, w_n, … are sequences with limits $lim\ u_n = u$, $lim\ v_n = v$, $lim\ w_n = w$, … then a well-behaved function $f(u_n, v_n, w_n, \ldots)$ converges to $f(u, v, w, \ldots)$. Brewer argues that since $t(y)$, $t(x)$ are unbiased for Y, X writing $\pi_i = $ Inclusion-probability $\sum_{s \ni i} p(s)$ for i of U, the limiting value of design-expectation of t_g is

$$lim\, E_p(t_g) = E_p(t(y)) + \frac{\sum\limits_{1}^{N} y_i x_i Q_i \pi_i}{\sum\limits_{1}^{N} x_i^2 Q_i \pi_i} (X - E_p(t(x))) = E_p(t(y)) = Y.$$

So, t_g is Asymptotic Design Unbiased and hence Asymptotic Design Consistent for Y.

In their work, Chaudhuri & Pal (2005) applied the simulation-based criteria of Average Coefficient of Variation, Actual Coverage Proportion (or Percentage) and Average Length (of Confidence interval) in assessing the comparative performances of the various alternative estimators based on varying sampling schemes. Here Average Coefficient of Variation refers to average over $R = 1000$ replicated samples of the Coefficient of Variation of a statistic, average coverage probability, namely, the proportion of the $R = 1000$ replicated samples for which a calculated Confidence Interval covers the estimated value of a parameter, and Average Length (of Confidence interval) is the average over the $R = 1000$ replicated samples, the value of the length of a Confidence Interval for a parameter as calculated for the respective samples. Desirably, Average Coefficient of Variation and Average Length (of Confidence interval) should be small and Actual Coverage Proportion (or Percentage) should be close to the aimed at value of a Confidence Interval. In survey sampling, usually no sampling design affords the estimator for a parameter with the least value of the variance of the estimator for a fixed vector of variable-values. So, this vector is postulated to be a random vector leading to a model-based super-population approach. Consequently Y cannot be estimated but can only be predicted by a statistic estimating the model-based expected value $E_m(Y)$ of this Y. Hence the name "predictor" rather than an "estimator".

7.1.4 Chaudhuri, Bose & Dihidar's (2005) Approach

Their work gives a thoroughly comprehensive account of (1) How 2-stage varying probability sampling, for example that given by Rao, Hartley & Cochran (1962), often needs to be supplemented into an Adaptive sample introducing well-illustrated various kinds of neighbourhoods, clusters and networks, (2) how initial estimators using only the variable of a major interest may be fruitfully improved utilizing auxiliary informative variables postulating various kinds of models so as to employ suitable Generalized Regression estimators according to amenable situations, (3) how to keep a check on the increasing Adaptive sample sizes going beyond control on taking random samples without replacement from the network already at hand and (4) exhibiting empirical evidences citing live data as well as simulation-based numerical exercises. Chaudhuri, Bose & Dihidar (2005) along with Chaudhuri (2000) should be companion material for any serious practitioner adopting Adaptive schemes of sample selection and employing estimation procedures relevant to the correspondingly drawn samples at least with a pragmatic view of such situations.

7.1.5 Chaudhuri, Bose & Ghosh's (2005) Observations

These authors consider simultaneous estimation of survey population totals of several variables by stratified 2-stage sampling with varying selection-probabilities. They employ model assisted Generalized Regression estimators and model dependent empirical Bayes' estimators adducing real life data situations. With the help of illustrated applications they demonstrate the efficacy of adaptive sampling. They cite various situations inducing alterations in model postulations leading to several greg predictor/estimator with varying levels of accuracies. Their data choices do not vindicate competitivity of empirical Bayes estimator against the greg predictor in their synthetic versions.

7.1.6 Thompson (1992), Thompson & Seber (1996) and Seber & Salehi's (2013) Approaches

As has been frankly admitted throughout, we have been inspired by Professors Thompson and Seber to work on Network sampling and Adaptive sampling. Fortunately, Profs J.K. Ghosh, Mausumi Bose and three of my junior colleagues gave me considerable support in my efforts. Our main departure is the relative simplicity in approach from the conventional one.

In Network sampling for every Observation Unit labelled i reachable through its link to one or more Selection Unit labelled j and identifiable, we need to observe y_i, the variate-value of interest for i and the "multiplicity" m_i namely the number of Selection Units to which i is linked. After that we need to calculate $w_j = \sum\limits_{i \in A(j)} \dfrac{y_i}{m_i}$ for j's in a sample s of Selection Units with $A(j)$ as the set of Observation Units linked to the jth Selection Unit. The sample s needs to be chosen by any sampling scheme for which every Selection Unit and every pair of distinct Selection Units have positive inclusion-probability. Rest of the theory is simple. Only a precaution needs to be in operation if resources demand curtailing the volume of the actual survey material.

In case of Adaptive sampling in our format we need to do the following.

Anticipating most of the sampling units in the population to contain none or negligible values for the variable y of interest but a few of them in unknown locations to contain them substantially and with a tendency for a plentiful localization, we need Adaptive sampling mainly for enhancing information content. For this we need aptly define neighbourhoods, clusters and networks. For every unit i in a sample s we need to identify the cluster $A(i)$ to which it belongs, find the cardinality C_i of $A(i)$ and obtain $t_i = \dfrac{1}{C_i} \sum\limits_{j \in A(i)} y_j$.

The rest of the theory including the mechanism for its constraining in case the Adaptive sample $A(s)$ which is the set of networks of the units in s, turns out too big for coverage is quite simple as in case of Network sampling.

As opposed to those, the contents and approaches of Thompson, Seber and Salehi are tremendously more comprehensive.

Seber & Salehi (2013) talk about Adaptive Sampling allowing neighbourhood to be amenable to expansion without being kept pre-determined, if needed to capture more units or areas.

Adaptive allocation is again one of their concepts not attended to by us.

A most glaring diversity is their concept of exploiting the counting of repeated appearance of networks depending on the number of times the units in respective networks happen to be selected. Recognition of this enables them to apply Rao-Black wellization in increasing efficiency in estimation.

Our approach just starts with the distinct units in the sample and identifying their respective networks in the estimation process.

However, we must categorically announce that for an aspirant for mastery in Adaptive sampling, the appreciation of the works of Thompson, Seber & Salehi and of those of the numerous authors they refer to are an absolute must.

One point we may make nevertheless. Murthy's (1957) strategy was clearly explained by us in our monograph by Chaudhuri & Stenger (2005) and also in its earlier version in 1988, published respectively by Taylor & Francis/CRC and Marcel Dekker. But in our view its applicability is not quite promising. However, Seber & Salehi (2013) hold contrary views.

Appendix

An outline of the technique of Constrained Size Network Sampling and its use in estimation.

Let M = the total number of known "Selection Units". These are supposed to be linked to the entire set of an unknown and unidentified second category of units called "Observation Units". Let N be the unknown but true number of Observation Units.

Let A_j be the set of Observational Units linked to the jth Selection Unit, $j = 1,\ldots, M$ and

$m_i \equiv$ the number of Selection Units linked to the ith Observational Unit; of course, $i = 1,\ldots, N$.

Let s be a sample of m Selection Units drawn from the population U of M Selection Units labelled $j = 1,\ldots, M$.

Let y be a real-valued-variable of interest with values y_i for $i \equiv 1,\ldots, N$. Our immediate object is to suitably estimate the population total $Y = \sum_1^N y_i$. Let $w_j = \sum_{i \in A_j} \dfrac{y_i}{m_i}, j \in s.$

Then,

$$W = \sum_1^M \sum_{i \in A_j} \frac{y_i}{m_i} = \sum_1^N \frac{y_i}{m_i} \left(\sum_{j \mid A_j \ni i} 1 \right) = Y.$$

So, it is enough to devise a suitable estimator for W using $(s, w_j | j \in s)$ in our effort to estimate Y. This approach of estimating the total of the values defined on the set of Observation Units on establishing a 'link' with a well-defined set of Selection Units and selecting a sample of Selection Units is called 'Network Sampling' use to estimate a parameter defined on the population of Selection Units shown to equal a parameter of interest related to the Observation Units.

By way of illustration suppose, $p_j \left(0 < p_j < 1, \sum_1^M p_J = 1 \right)$ be available as M normed size measures for the Selection Units. Then, a sample of Selection Units from U may be selected on applying the celebrated scheme given by Rao, Hartley and Cochran (1962). For this m mutually exclusive random groups are formed out of all the Selection Units, taking, say M_j Selection Units in the jth group $(j = 1, \ldots, m)$ such that $\sum_m M_j = M$, writing $\sum_m$ to denote sum over the m groups. Optimal group sizes M_j's are recommended by Rao, Hartley and Cochran (1962) as:

$$M_j = \left[\frac{M}{m} \right], j = 1, \ldots, K$$

$$= \left[\frac{M}{m} \right] + 1, j = K + 1, \ldots, m.$$

Here K is so determined uniquely to yield

$$\sum_m M_j = M .$$

Writing $Q_j \equiv$ the sum of the values of p_j's for the Selection Units falling in the jth group, $j = 1, \ldots, m$ it follows that

$$t = \sum_m \left(\frac{Q_j}{p_j} \right) w_j$$

is an unbiased estimator for $W = Y$.

Writing $\sum\sum\limits_{M\ M}$ the sum of the pairs of distinct units with no repetition, from Rao-Hartley-Cochran (1962), we see that the variance of this estimator t is

$$V(t) = A \sum_{M}\sum_{M} P_j P_{j'} \left(\frac{w_j}{p_j} - \frac{w_{j'}}{p_{j'}} \right)^2$$

with

$$A \frac{\sum\limits_{m} M_j^2 - M}{M(M-1)}.$$

Writing $\sum\sum\limits_{m\ m}$ as the sum of the pairs of distinct groups with no repetition they have given a uniformly non-negative unbiased estimator of that variance $V(t)$ as

$$v(t) = B \sum_{m}\sum_{m} Q_j Q_{j'} \left(\frac{w_j}{p_j} - \frac{w_{j'}}{p_{j'}} \right)^2$$

writing

$$B \frac{\sum\limits_{m} M_j^2 - M}{M^2 - \sum\limits_{m} M_j^2}.$$

If s be taken as a Simple Random Sample Without Replacement of size m, then, an unbiased estimator for $W = Y$ may be taken as

$$t' = \frac{M}{m} \sum_{j \in s} w_j.$$

Its variance is

$$V(t') = M^2 \left(\frac{1}{m} - \frac{1}{M} \right) \frac{1}{(M-1)} \sum_{j=1}^{M} \left[w_j - \frac{1}{M} \sum_{j}^{M} w_j \right]^2$$

An unbiased estimator of this $V(t')$ is

$$v(t') = M^2 \left(\frac{1}{m} - \frac{1}{M} \right) \frac{1}{(m-1)} \sum_{j \in s} \left[w_j - \frac{\sum\limits_{j \in s} w_j}{m} \right]^2$$

In the respective cases the accuracy of an estimator e for $W = Y$, say, $e = t, t'$ will be measured by

$$CV = 100 \times \frac{\sqrt{v(e)}}{e}$$

Whatever may be a sample s of the Selection Units taken, the corresponding "Network Sample", namely the sample of Observation Units linked to the Selection Units in s, may be quite large. As these Observation Units are actually to be surveyed, often with limited budget and resources, it may be hard to execute the survey of the Observation Units in the 'Network Sample'. So, we may actually survey the "Constrained Network Sample", derived as follows:

Let $C_j \equiv$ the number of Observation Units in A_j. Then, $C = \sum\limits_{j \in s} C_j$

is the total number of Observation Units to be surveyed. If C be

prohibitively large, let from the respective A_j's independently across j in s, Simple Random Samples Without Replacements of sizes $d_j(2 \le d_j \le C_j)$ be taken as B_j's such that $D = \sum\limits_{j \in s} d_j$ is kept within a

manageable number. This set, namely, $S_B = U_{j \in s} B_j$ is our "Constrained Network Sample". Then the estimator t is to be modified as

$$u = \sum_{j \in s} \frac{Q_j}{P_j} \left(\frac{C_j}{d_j} \sum_{i \in B_j} \frac{y_i}{m_i} \right) = \sum_{j \in s} \frac{Q_j}{P_j} u_j, \text{ say,}$$

writing

$$u_j = \left(\frac{C_j}{d_j} \right) \sum_{i \in B_j} \left(\frac{y_i}{m_i} \right).$$

Then, with some "none too difficult" algebra, an unbiased estimator of the variance of u turns out as

$$v(u) = (1+B)\sum_m v_L(u_j)\left(\frac{Q_j}{P_j}\right)^2 + B\left[\sum_{j \in s}(u_j)^2 \frac{Q_j}{P_j} - u^2\right]$$

writing

$$v_L(u_j) = C_j^2\left(\frac{1}{d_j} - \frac{1}{C_j}\right)\left(\frac{1}{d_j-1}\right)\sum_{i \in B_j}\left(\frac{y_i}{m_i} - \frac{\displaystyle\sum_{i \in B_j}\frac{y_i}{m_i}}{d_j}\right)^2$$

Similarly, t' is to be modified into $\hat{t} = \left(\dfrac{M}{m}\right)\displaystyle\sum_{j \in s} u_j$ while employing a 'Constrained Network Sample'. Then, an unbiased estimator of variance of $\hat{t}$ turns out as

$$v(\hat{t}) = \left(\frac{M}{m}\right)^2\left[\sum_{j \in s} v_L(u_j) + \left(\frac{M}{m-1}\right)\left(\frac{1}{m} - \frac{1}{M}\right)\sum_{i<j \in s}(u_i - u_j)^2\right]$$

Further details need not be shown.

Bibliography

Basu, D. (1958). On sampling with and without replacement. *Sankhyā*, 20, 287–94.

Bharati, P., Pal, M., Vasulu, T.S., Bhattacharya, B., Chaudhuri, A., Das, R.N., Datta, T. and Bose, K. (1998). Estimation of infant and maternal mortality rates and identification of their determinants in North 24 Parganas, West Bengal. *A Unicef-sponsored Project Report*, Indian Statistical Institute Kolkata.

Bharati, P., Bharati, S., Pal, M., Chakraborty, S. and Gupta, R. (2008). Chronic energy deficiency among Indian women by residential status. *Ecology of Food and Nutrition*. Routledge/Taylor & Francis. Boca Raton, Florida, USA.

Basu, D. and Ghosh, J.K. (1967). Sufficient statistics in sampling from a finite universe. *Bull. Int. Stat. Inst.* 36, 850–59.

Brewer, K.R.W. and Hanif, M. (1983). Sampling with unequal probabilities: *Lecture Notes in Statistics,* No.15. Springer Verlag. New York, USA.

Brown, J.A. (1994). The application of adaptive cluster sampling to ecological studies. In *Statistics in Ecological and Environmental Monitoring*, 2nd ed., D.J. Fletcher and B.F.J. Manly (Eds). Dunedin, New Zealand University of Otago Press, pp 86–97.

Cassel, C.M., Särndal, C.E. and Wretman, J.H. (1976). Some results on generalized difference estimation and generalized regression estimation for finite population. *Biometrika*, 63, 615–20.

Cassel, C.M., Särndal, C.E. and Wretman, J.H. (1977). John Wiley & Sons., New York, USA.

Chaudhuri, A. (2000). Network and adaptive sampling with unequal probabilities. *Cal. Stat. Assoc. Bull.*, 50, 238–53.

Chaudhuri, A. (2003). Estimation from an under-covered sample in a complex survey for auditing. *Cal. Stat. Assoc. Bull.*, 54, 115–20.

Chaudhuri, A. (2010). *Essentials of Survey Sampling*. PHI Learning, Delhi.

Chaudhuri, A. (2011). Survey on Drug Abuse as summarized in (I) National Household Survey of Drug and Alcohol Abuse in India by Prof. Anurag Srivastava as Principal Investigator, 2003, (II) Feasibility of Conducting Pilot Survey on Drug Abuse— Report of the Group, Ministry of Statistics and Programme Implementation, Govt. of India, November, 2007 and (III) Report of the Committee to suggest appropriate survey methodology for undertaking Pilot Survey on Drug Abuse, NSSO, MOSPI, GOI, March 2009.

Chaudhuri, A. and Dihidar, S. (2010). How to constrain explosive size of a Network Sample?

————— an unpublished Report of a survey in Assam on Household Survey expenses for treatment of inmates for heart, cancer and gall bladder-related diseases as clinical in-patients, sponsored by Indian Statistical Institute, 2006.

Chaudhuri, A. and Pal, S. (2002). On certain alternative mean square error estimators in complex survey sampling. *Jour. Stat. Plan. Inf.*, 104, 363–75.

Chaudhuri, A. and Pal, S. (2005). Estimating domain-wise distribution of scarce objects by adaptive sampling and model-based borrowing of strength. *Jour. Ind. Soc-Agri Stat.*, 58(I), 136–43.

Chaudhuri, A. and Saha, A. (2004). On adaptive cluster sampling and generalized regression method of estimating localized elements. *Jour. Prob. Stat. Sc.*, 2(1), 35–45.

Chaudhuri, A. and Stenger, H. (2005). *Survey Sampling: Theory and Methods*, 2nd ed., Chapman & Hall/CRC Taylor & Francis Group. Boca Raton, Florida, USA.

Chaudhuri, A. and Vos, J.W.E. (1988). *Unified Theory and Strategies of Survey Sampling*. North-Holland Amsterdam, Netherlands.

Chaudhuri, A., Adhikary, A.K. and Dihidar, S. (2000). Mean square error estimation in multi-stage sampling. *Metrika.*, 52, 115–31.

Chaudhuri, A., Bose, M. and Dihidar, K. (2005). Sample-size restrictive adaptive sampling: An application in estimating localized elements. *Jour. Stat. Plan. Inf.*, 134, 254–67.

Chaudhuri, A., Bose, M. and Ghosh, J.K. (2004). An application of adaptive sampling to estimate highly localized population segments. *Jour. Stat. Plan. Inf.*, 121, 175–89.

Chaudhuri, A., Bose, M. and Ghosh J.K. (2005). Improved simultaneous estimation of small domain sizes through model assistance. *Jour. Prob. Stat. Sc.*, 3(1), 17-29.

Cochran, W.G. (1953). *Sampling Techniques.* John Wiley & Sons., New York, USA.

Cochran, W.G. (1963). *Sampling Techniques*, 2nd ed., John Wiley & Sons., New York, USA.

Cochran, W.G. (1977). *Sampling Techniques*, 3rd ed., John Wiley & Sons., New York, USA.

Cramér, H. (1946). *Mathematical Methods of Statistics.* Princeton University Press.

Freund, J.E. (1994). *Mathematical Statistics*, 5th ed., Prentice Hall International. Englewood Cliffs, N.J., USA.

Goodman, L.H. (1961). Snow-bak Sampling. *Ann. Math. Stat.* 32, 148–70.

Hartley, H.O. and Ross, A. (1954). Unbiased ratio estimators. *Nature*, 174, 270–71.

Ha'jek, J. (1959). Optimum strategy and other problems in probability sampling. *Cas. Pest. Mat.*, 84, 387–473.

Hansen, M.H. and Hurwitz, W.N. (1943). On the theory of sampling from finite populations. *Ann. Math. Stat.* 14, 333–62.

Hanurav, T.V. (1966). Some aspects of unified sampling theory. *Sankhyā*, A . 28, 175-204.

Hege, V.S. (1965). Sampling designs which admit uniformly minimum variance unbiased estimators. *Cal. Stat. Assoc. Bull.*, 14, 160–62.

Horvitz, D.G. and Thompson, D.J. (1952). A generalization of sampling without replacement from a finite universe. *Jour. Amer. Stat. Assoc.*, 47, 663–85.

Lahiri, D.B. (1951). A method of sample selection providing unbiased ratio estimators. *Bull. Int. Stat. Inst.*, 33(z), 133–40.

Lanke, J. (1975). Some contributions to the theory of survey sampling. Unpublished but widely circulated Ph.D. thesis. Univ. Lund. Sweden.

Midzuno, H. (1952). On the sampling system with probabilities proportionate to the sum of sizes. *Ann. Inst. Stat. Math.*, 3, 99–107.

Murthy, M.N. (1957). Ordered and unordered estimators in sampling without replacement. *Sankhyā*, 18, 379–90.

Pal, M., Bharati, P., Vasulu, T.S., Chaudhuri, A., Bhattacharya, B., Das, R.N. and Ghosh, R. (2002). Maternal Mortality in Rural North 24 Parganas, West Bengal: Estimation of its Rate and Identification of underlying causes. *Demography India*. 31, 253–58.

Prasad, N.G.N. and Rao, J.N.K. (1990). The estimation of mean squared errors of small-area estimators. *Jour. Amer. Stat. Assoc.*, 85, 163–71,

Rao, J.N.K. (1979). On deriving mean square errors and other non-negative unbiased estimators in finite population sampling. *J. Amer. Stat. Assoc.*, 17, 125–36.

Rao, J.N.K. (1999). Some current trends in sample survey theory and methods. *Sankhyā* B, 61, 1-57.

Rao, J.N.K., Hartley, H.O., and Cochran, W.G., (1962). On a simple, procedure of unequal probability sampling without replacement. *Jour. Roy. Stat. SOC. Scr.*, B24, 482–91.

Salehi, M.M. and Seber, G.A.F. (1997). Adaptive cluster sampling with networks selected without replacement. *Biometrika*. 84(1), 209–19.

Salehi, M.M. and Seber, G.A.F. (2002).Unbiased estimators for restricted adaptive cluster sampling. Aust & New Zealand, *J. Stat.*, 44(1), 63–74.

Seber, G.A.F. and Salehi, M.M. (2013). *Adaptive Sampling Designs. Inference for Sparse and Clustered Populations*. Springer. New York, Heidelberg.

Sen, A.R. (1953). On the estimator of the variance in sampling with varying probabilities, *Jour. Ind. SOC. Agri.*, 5(2), 119–27.

Sirken, M.G. (1970). Household surveys with multiplicity. *Jour. Amer. Stat. Assoc.*, 65, 259–66.

Sirken, M.G. (1983). Handling missing data by network sampling. In Incomplete data in Survey Samling. 2, 81-90.

Thompson, S.K. (1990). Adaptive cluster sampling. *Jour. Amer. Stat. Assoc.*, 85, 1050–59.

Thompson, S.K. (1992). *Sampling*, Wiley. New York, USA.

Thompson, S.K. and Seber, G.A.F. (1996). *Adaptive Sampling.* Wiley, New York, USA.

Yates, F. and Grundy, P.M. (1953). Selection without replacement from within strata with probability proportional to size. *Jour. Roy. Stat. SOC. Scr.*, B15, 253–61.

Author Index

Subject Index